BIOMEDICAL ETHICS REVIEWS

Edited by ***James M. Humber and Robert F. Almeder***

BIOMEDICAL ETHICS REVIEWS

Edited by *James M. Humber and Robert F. Almeder*

Mental Illness and Public Health Care • 2002

Privacy and Health Care • 2001

Is There a Duty to Die? • 2000

Human Cloning • 1999

Alternative Medicine and Ethics • 1998

What Is Disease? • 1997

Reproduction, Technology, and Rights • 1996

Allocating Health Care Resources • 1995

Physician-Assisted Death • 1994

Bioethics and the Military • 1992

Bioethics and the Fetus • 1991

Biomedical Ethics Reviews • 1990

Should the U.S. Adopt a National Health Insurance Plan? · Are the NIH Guidelines Adequate for the Care and Protection of Laboratory Animals?

Biomedical Ethics Reviews • 1989

Should Abnormal Fetuses Be Brought to Term for the Sole Purpose of Providing Infant Transplant Organs? · Should Human Death Be Taken to Occur When Persons Permanently Lose Consciousness?

Aids and Ethics • 1988

Biomedical Ethics Reviews • 1987

Prescribing Drugs for the Aged and Dying · Animals as a Source of Human Transplant Organs · The Nurse's Role: *Rights and Responsibilities*

Quantitative Risk Assessment: *The Practitioner's Viewpoint* • 1986

Biomedical Ethics Reviews • 1985

Buying and Selling of Human Organs · Sex Preselection · Medical Decisionmaking Under Uncertainty · Concepts of Health and Disease

Biomedical Ethics Reviews • 1984

Public Policy and Research with Human Subjects · The Right to Health Care in a Democratic Society · Genetic Screening · Occupational Health The Ethics of Fetal Research and Therapy

Biomedical Ethics Reviews • 1983

Euthanasia Surrogate Gestation · The Distribution of Health Care The Involuntary Commitment and Treatment of Mentally Ill Persons Patenting New Life Forms

Mental Illness and Public Health Care

Edited by

James M. Humber

and

Robert F. Almeder

Georgia State University, Atlanta, Georgia

Humana Press • Totowa, New Jersey

999 Riverview Drive, Suite 208
Totowa, NJ 07512

humanapress.com   email: humana@humanapr.com

For additional copies, pricing for bulk purchases, and/or information about other Humana titles, contact Humana at the above address or at any of the following numbers: Tel.: 973-256-1699; Fax: 973-256-8341; E-mail: humana@ humanapr.com, or visit our Website: humanapress.com

This publication is printed on acid-free paper. ∞

ANSI Z39.48-1984 (American National Standards Institute) Permanence of Paper for Printed Library Materials.

Cover design by Patricia F. Cleary.
Production Editor: Jessica Jannicelli.

ISBN 1-58829-021-2

Printed in the United States of America. 10 9 8 7 6 5 4 3 2 1

The Library of Congress has cataloged this serial title as follows:

Biomedical ethics reviews—1983– Totowa, NJ: Humana Press, c1982–
v.; 25 cm—(Contemporary issues in biomedicine, ethics, and society)
Annual.
Editors: James M. Humber and Robert F. Almeder.
ISSN 0742-1796 = Biomedical ethics reviews.
1. Medical ethics—Periodicals. I. Humber, James M. II. Almeder, Robert F.
III. Series.
[DNLM: Ethics, Medical—periodicals. W1 B615 (P)]
R724.B493   174'.2'05—dc19   84-640015
AACR2 MARC-S

Contents

Preface

The articles in this latest volume of *Biomedical Ethics Reviews* focus on three specific issues relative to the general topic heading, *Mental Illness and Public Health Care*. The first of these issues is whether or not the involuntary commitment of mentally ill persons can be said to be morally proper, or even morally permissible, in a society such as ours. The questions arising in connection with this issue are complex. For example, is dangerousness to oneself or others a sufficient ground for committing a mentally ill person to an institution contrary to their will? Are mental health professionals competent to predict dangerousness? Committing a person to an institution for their own good is paternalistic; can this paternalism be justified in a liberal society? In the first two essays in this text, Theodore Benditt and Gerard Elfstrom attempt to answer questions such as these. Although their approaches differ radically, the reader will find that they nevertheless come to quite similar conclusions.

The second topic of discussion in our text is a very broad one: How should we go about determining proper psychiatric care within the parameters of our present health care delivery system? Three articles are devoted to this issue. In the first essay, David Malloy and Thomas Hadjistavropoulos argue that whenever the use of cognitive behavioral therapy (CBT) and pharmacological interventions are both in accord with professional codes of conduct and approximately equal in terms of their effectiveness, CBT should be the treatment of choice because it possesses an ethical advantage. In the essay that follows, Mark Meaney argues that publicly funded managed care for behavioral health services can efficiently and effectively provide benefits for patients. What is needed, he says, is for health services to take care to integrate ethics into their operations. To show how this can be done, Meaney examines an actual case in which a Philadelphia-based public sector managed behavioral health care corporation used the services of an ethics center in Atlanta to implement a system-wide corporate ethics program. In the final essay, Wade Robison expresses concern that our current system of psychiatric care exhibits a shift away from the traditional Freudian model of open-ended, one-on-one therapy to fewer doctor–patient consultations combined with greater use of pharmacological treatments. Robison acknowledges that this transformation in psychiatric care may have some benefits, but he is extremely troubled by the fact that the principal driving forces behind the change in treatment are economic rather than deliberations concerning what forms of treatment will most benefit patients.

The last issue in our text concerns what should be done when a mental health professional is convinced that one of his or her patients poses a threat to someone else in society. Here the primary concern is whether psychotherapists should break patient confidentiality and warn those whom they believe are in jeopardy at their patients' hands. In 1976, in the case of Tarasoff v. Regents of the University of California, the California Supreme Court determined that psychotherapists have a duty of reasonable care to protect those whom they believe could be harmed by a patient. This decision has, in one form or another, been incorporated into most states' laws. In "Tarasoff, Megan, and Mill: *Preventing Harm to Others*," Pam Sailors argues that these laws contain some deficiencies and that the cure for these deficiencies is to modify Tarasoff laws so that they all more closely resemble "Megan's Law"—a law that requires various law enforcement agencies to release relevant information in an attempt to protect the public from sexual offenders.

Mental Illness and Public Health Care is the nineteenth annual volume of *Biomedical Ethics Reviews*, a series of texts designed to review and update the literature on issues of central importance in bioethics today. For the convenience of our readers, each article in every volume of our series is prefaced by a short abstract describing that article's content. Each volume in the series is organized around a central theme; the theme for the next issue of *Biomedical Ethics Reviews* will be *Care of the Aged.* We hope our readers will find the present volume of *Biomedical Ethics Reviews* to be both enjoyable and informative, and that they will look forward with anticipation to future volumes on topics of special concern.

James M. Humber
Robert F. Almeder

Contributors

Theodore Benditt • Department of Philosophy, University of Alabama at Birmingham, Birmingham, Alabama

Gerard Elfstrom • Department of Philosophy, Auburn University, Auburn, Alabama

Thomas Hadjistavropoulos • Department of Psychology, University of Regina, Regina, Canada

David Cruise Malloy • Faculty of Kinesiology and Health Studies, University of Regina, Regina, Canada

Mark E. Meaney • Midwest Bioethics Center, Kansas City, Missouri

Wade L. Robison • Department of Philosophy, Rochester Institute of Technology, Rochester, New York

Pam R. Sailors • Department of Philosophy, Southwest Missouri State University, Springfield, Missouri

Abstract

A central question regarding mental illness and civil commitment is whether a person who is not violating any laws may be involuntarily committed and treated. After a lengthy period of time during which very erratic behavior was regarded as sufficient, civil liberties concerns led to the tightening of grounds for commitment and treatment; dangerousness to self and/or to others became the standard. The consequent appearance in our communities, however, of noticeable numbers of apparently mentally ill people produced a reaction: Many people came to believe that the good of such individuals, our capacity to treat them effectively, should be the standard, whether those being treated wanted it or not.

This chapter tries to mediate the conflict between civil libertarian and paternalistic concerns, between people's rights and our desire to help. The chapter tests our thoughts about the issues by exploring a continuum of hypothetical cases that vary, on the one hand, the degrees to which individuals are harmed by their mental illnesses and, on the other, the extent to which their behavior imposes on others. The chapter's conclusion is that if we are going to go beyond dangerousness as the standard, we need to ground our interference on behavior we are willing to proscribe. If we are not willing to make certain behavior legally unacceptable, we cannot use it as a basis for committing someone to an institution.

The chapter also takes issue with the contention that there really is no such thing as mental illness because "real" illness involves a detectable underlying pathology. It is argued, to the contrary, that the medical profession regularly, and correctly, deals with conditions in which this is not the case.

Mental Illness and Commitment

Theodore Benditt

"... many people ... are profoundly ambivalent in their attitudes toward the mentally ill. We are frightened by them and we seek distance and protection from them, yet we also feel compassion toward them and we urge sympathetic care and protection of them."[1]

"In the end it is up to society to determine which statistically abnormal states and conditions to regard as illnesses according to the assessment of desirable human experiences and functioning."[2]

"Involuntary mental hospitalization is imprisonment under the guise of treatment; it is a covert form of social control. . . ."[3]

"... schizophrenia is a housing problem rather than a medical problem. . . ."[4]

"To solve the problem of mentally ill street people, we are going to have to revive some form of mental asylum."[5]

Over the past several decades, attitudes toward mental illness and the way we deal with it have undergone change, from

From: *Biomedical Ethics Reviews: Mental Illness and Public Health Care*
Edited by: J. Humber & R. Almeder © Humana Press Inc., Totowa, NJ

great confidence in the capacity of psychiatry to respond to mental problems, to doubts about this and about the propriety of forcing treatment on those thought to need it, to a resurgence of belief in psychiatry's ability to help people and in civil commitment. The first of these alterations came in the context of the civil rights movement in the United States, which was extended, through judicial action, to the rights of those seen as mentally ill. The reaction reflected both a concern with the consequences of the civil libertarian approach to mental illness—a concern both for those seen to be mentally ill and for the larger society—and the belief that new drugs could cope with some of the most serious mental illnesses.

Accordingly, many issues in mental illness are contests between those having more and those have less expansive views about what mental illness is, our capacity to treat it, and the propriety of doing so against the wishes of those held to be mentally ill. On one side are those who think that many people suffer from mental illnesses, that we have the ability to treat them, that treatments, though they may not (always) cure, at least improve people's lives, and that mentally ill people have a right to treatment even when they cannot understand that they will benefit from it. These views are contested by those who think that the problems are not genuine medical problems but rather problems in living, that most people with so-called mental illnesses are making choices for which they should be held responsible, that involuntary commitment is simply a form of social control that is not a legitimate part of medical practice, and that the right to liberty is violated by involuntary commitment and treatment.

Is There Such a Thing as Mental Illness?

Over the course of 40 years, the case against the very idea of mental illness has been strenuously argued by Dr. Thomas Szasz, a psychiatrist. He believes that so-called mental illness does not genuinely qualify as illness, that the behavioral problems associated with so-called mental illnesses should be dealt with, if at all, through various forms of social control but not by medicine or psychiatry, and that the use of civil commitment in dealing with the so-called mentally ill turns psychiatrists into agents of the state for the control of certain people. Many have credited Szasz (not always approvingly) with having had a considerable impact on the reduction of the number of people in mental institutions and on the procedures used to commit and to treat the mentally ill.

Most of the psychiatric profession, which is more and more pharmacologically based, is arrayed against Szasz's views as to the nature of mental illness and the propriety of civil commitment. Taken as a whole, the profession maintains that many mental illnesses are genuine even in Szasz's sense (i.e., that they are brain diseases) and that they are treatable. The American Psychiatric Association publishes a massive volume entitled *Diagnostic and Statistical Manual of Mental Disorders* (usually referred to as DSM, it is now in its fourth edition) in which hundreds of mental disorders and the bases for diagnosing them are identified. For his part, Szasz regards many of these not as illnesses, but as problems in living that are not the province of medicine. However, although one can sympathize with the worry that the extravagant list of disorders in DSM-IV inflates and perhaps diminishes the idea of mental illness,[6] it is hard not to accept that there are genuine mental illnesses, although Szasz continues to reject it. In a recent interview, he was asked the following:

> The psychiatrist E. Fuller Torrey has written that "studies using techniques such as magnetic resonance imaging and positron emission tomography scans have proved that schizophrenia and manic depressive illness are physical disorders of the brain in exactly the same way as Parkinson's disease or multiple sclerosis." Is that true? If not, what do these studies actually show?

Szasz answered:

> Most educated people, if they think about it, know how real disease is diagnosed. Take anemia. If a person comes

> in and says he is tired, he has no energy, and he looks very pale, the physician may think he is anemic. But the diagnosis is not made until there is a finding in the laboratory that there is a diminished blood count, a diminished hemoglobin level. Conversely, a laboratory technician can blindly make a diagnosis of anemia simply on the basis of vials of blood submitted to him or her—without having any idea of whose blood it is. As soon as that can be done with schizophrenia, it will be a brain disease, exactly as neurosyphilis was recognized as a brain disease.[7]

By Szasz's account, mental disease (or illness) is not real disease (or illness). In a real illness, he says, we can distinguish between outward manifestations or occurrences (lack of energy, pale looks, tiredness) and an underlying problem, an underlying physiological state of affairs, an organic pathology (such as diminished hemoglobin level) that is responsible for the outward manifestations. With respect to most so-called mental illnesses, however, Szasz says this distinction does not hold—there are certainly outward occurrences, but in most cases there is no organic pathology that is the mark of real illness. For this reason, he says, most so-called mental illness is not a medical problem. Szasz identifies illness with the underlying pathological situation,[8] which is what medical treatment is typically directed toward correcting.

People seek medical attention when they experience things (what I have been calling outward manifestations or occurrences) that are uncomfortable and/or out of the ordinary and/or worrisome. A physician makes a diagnosis. Sometimes the diagnosis reveals an underlying state of affairs that is regarded as a disease or illness; sometimes it does not. What makes the underlying pathology an *illness* is that it not only is the cause of the outward manifestations but also poses a threat to future health and well-being. In such a case, the outward occurrences are *symptoms* of the illness. Another way of putting Szasz's contention, then, is that, in genuine illness, the outward occurrences that lead people to seek medical attention are *symptoms* of an underlying problem that has an existence and can be identified independent of them. (Indeed, according to Szasz, an illness can exist and be identified independent of symptoms—in principle, there need not be any symptoms when the illness is discovered.) By this criterion, Szasz believes, most so-called mental illness is not illness: The outward manifestations are not symptoms. Despite this, though, the public and the psychiatric profession insist, with no foundation (as Szasz sees it), that so-called mental illnesses really are illnesses or diseases—so-called *brain* diseases.[9]

Szasz overstates his case, though—for there are many sorts of legitimately medical situations in which outward manifestations are not symptoms of an underlying pathology posing a threat to future health and well-being. Consider something like a rash. Although a rash might be both caused by and a symptom of an underlying illness, it need not be. It might be a local or transitory occurrence, the product of an irritation or a passing internal state of affairs that does not qualify as disease or illness (as that has been characterized earlier). Although it is *caused by* some state of the body, it need not be a *symptom of* an illness. So too with mental and emotional difficulties.

Contrary to what Szasz implies, medicine is not limited to dealing with illnesses in the strict sense on which his argument depends. Medicine also deals with outward occurrences. For example, physicians deal with broken limbs, rashes, and a variety of aches and pains. They try to ascertain whether there is anything going on that portends further difficulty. If there is, they try to deal with it, although dealing with it may mean only monitoring it as it runs its course or corrects itself. Frequently, however, they respond to the outward occurrences quite independently of whether there is known to be (or even is at all) a problematic pathology underlying it.

Much mental illness is in certain respects like a rash, at least at the present state of our knowledge (although the comparison is not meant to diminish the seriousness of mental illness). Certain outward occurrences are presented which are thought to be problematic either by the person manifesting them or by others. They are (we believe) in some way caused by the brain and nervous system. Whether they are *symptoms* of a problematic underlying pathology remains to be seen—some are and others are not (or are not yet known to be). However, even when they are not, there is no reason not to see them as medical issues, just as we see rashes and broken limbs as medical issues, for the medical profession deals with many outward occurrences that may betoken nothing beyond themselves. There is this difference between rashes and broken limbs, on the one hand, and many mental illnesses: The former are disorders of the body, whereas mental illness is often behavioral, affective, and cognitive. However, this in itself is no ground for denying medicine a role, if medicine, in fact, has any capacity to deal with these difficulties.

Civil Commitment: A Historical Sketch

Having said that there are genuine mental illnesses and a legitimate medical role for the psychiatric profession (however pharmacological it has become) only resolves a threshold issue with respect to the problem of civil commitment, for the psychiatric profession as a whole believes that certain individuals need, and indeed have a right to, treatment and that, therefore, civil commitment is needed and justified.

There was a time in our history when civil commitment was not the problematic issue it has become. For a long time, our society has felt a need to "deal" with the mentally ill. Either for their own good or to relieve families of the burden of dealing with mentally ill family members, states in the United States established institutions for the long-term housing and care of the mentally ill. Once called asylums, or insane asylums, they later came to be called state hospitals, or state mental hospitals. People were typically sent to these institutions at the instigation of their families or the police, on the basis of the affirmation of physicians that they needed treatment and were likely to benefit from it. Although minimal hearings were frequently required for making these determinations, such procedures were sometimes deemed too slow and cumbersome and expedited commitment procedures were often used.

This was the state of affairs up to the 1960s when concern began to grow, fueled in part by Szasz's writings, about the reality of mental illness, the capacity of the medical profession to deal with it, and the effect of institutionalization on those committed to the state hospitals. Questions about civil rights and the legitimate reach of state power that were being asked in other areas of social life came to be asked about the treatment of the mentally ill, and as in other areas, these questions found their way into the courts. In *Lessard v. Schmidt* (1972), a federal district court in Wisconsin held that in a civil commitment hearing, it must be shown that the mentally ill person risks causing immediate harm to others or to himself and also held that more rigorous, criminal-style, procedural safeguards are required.[10] "*Lessard* came to symbolize a desire to subordinate the therapeutic impulse that once drove civil commitment law to the civil libertarian concerns that were then in ascendance."[11] At the same time, another series of cases, beginning with *Wyatt v. Stickney* (1972)[12] in Alabama and *O'Connor v. Donaldson* (1975),[13] arising in Florida, held that committed mental patients could not be kept in mental hospitals if they were not being treated—whether because of insufficient resources or because no useful treatment was known. Finally, yet another series of decisions held that mental patients, even though properly committed, may refuse treatment and may be treated against their will only following a judicial-like inquiry into their competence.

The upshot of all of these civil liberties rulings and of the advent of new psychotropic drugs was a massive "deinstitutionalization"—mental patients were turned out of state hospitals to be dealt with instead by community mental health facilities, a move that was initially welcomed by many psychiatrists.[14] Over time, however, many in the mental health professions came to have second thoughts about the changes. Community mental health facilities did not materialize; many people deemed to be mentally ill wound up in jails or on the streets (according to some, deinstitutionalization had become merely "transinstitutionalization"); their presence in our cities became noticeable. Among many, both lay and professional, the belief took hold that something could and should be done for these people (even if they themselves did not want treatment), that people were "dying with their rights on."[15] By the early 1980s, the American Psychiatric Association proposed a basis for commitment that focused more on the need for treatment than on dangerousness (commitment would be allowed if the patient "will if not treated suffer or continue to suffer severe and abnormal mental, emotional, or physical distress, and this distress is associated with significant impairment of judgment, reason, or behavior causing a substantial deterioration of his previous ability to function on his own").[16]

Joyce Brown

The case of Joyce Brown is a good point of reference for conflicting ideas about civil commitment, for it raises both the question of whether someone is mentally ill and the difficult issue of the intersection of mental illness and homelessness. In response to growing numbers of homeless mentally ill people, New York City established a program called Project Help to determine whether such people could benefit from psychiatric treatment. At first, the program was voluntary except for those who were dangerous to themselves or others. In 1987, under then-Mayor Ed Koch, the criteria were expanded to include "need to be treated" and "self-neglect." Using the new criteria, Project Help picked up Joyce Brown. Mayor Koch was aware of this woman's situation, believed she could be treated, and wanted to help her (critics attributed other motives to the mayor). Joyce Brown lived on a steam grate on New York City's Upper East Side. She took money from passersby, with which she bought food, although sometimes she tore up the money. She talked to people who passed by, often shouting obscenities. She urinated on the side-walk and defecated in the gutter and (some claimed, although she denied it) on herself. She sometimes ran into the street where there was traffic. She often wore little clothing despite cold weather, although she kept sheets and blankets in which to wrap herself when she slept.

At a commitment hearing after she was picked up, Joyce Brown said she was a "professional street person." Her American Civil Liberties Union (ACLU) attorneys said she did not want help, did not need help, and was entitled to live as she pleased. They said her homelessness was a matter of choice and that, in that context, her behavior was rational. The outcome was that the judge declined to honor Project Help's new criteria and ruled that Joyce Brown was not a danger to herself or others. This ruling was overturned on appeal, resulting in her commitment in Bellevue Hospital. She refused treatment, however, thus triggering another hearing at which the court ruled she was competent and could not be medicated without consent—so the hospital released her. Later, Brown said "The only thing wrong with me was that I was homeless, not insane. . . . I need a place to live; I don't need an institution. . . ."[17] For a while after her release she lived in a hotel room and looked for a job. She became a celebrity of sorts, appearing on talk shows and addressing a class at Harvard Law School. After a while, she was seen back on the streets; later she lived in a residence setting, was in and out of hospitals, and had problems with illegal drugs.[18]

Widespread knowledge of the Joyce Brown case makes it particularly useful in discussing principles governing civil commitment; I will use it to discuss bases for civil commitment *other* than dangerousness. Many other issues are raised by the case that cannot be pursued here: both the general problem of homelessness and the part of the homelessness problem that can be traced to deinstitutionalization; the role of illegal drugs as they relate either to mental illness or homelessness; the legal criteria (dangerousness to self or others) that dominate discussion at the present time. What *will* be discussed is whether there is a suitable basis for civil commitment *other* than dangerousness.[19]

The central conundrum of civil commitment is: For whose good are people being coerced on the ground that they are mentally ill—theirs or ours? Although some questioned his motives, Mayor Koch insisted that he wanted to help Joyce Brown. A great many commentators, claiming motives similar to Mayor Koch's, have expressed concern that the dangerousness standard is overprotective and should be replaced by need-for-treatment criteria along the lines of the American Psychiatric Association's model law, mentioned earlier. Thomas Szasz, on the other hand, says

> Joyce Brown was not demented; she knew what she was doing and, as a reward for her exploits, was invited to lecture at Harvard Law School. . . . Brown was not committed "for her own safety," but for the benefit of the community. . . . [T]he Brown case . . . illustrates our collective enthusiasm for avoiding the use of the criminal justice system as a means of controlling a large class of lawbreakers. . . . [P]rotecting liberty and property from those who disrespect or destroy them ought to be the task of judges, juries, and prison guards, not psychiatrists, psychologists, and social workers. And the means of enforcing such protection should be the criminal justice system, not the mental health system.[20]

What I want to propose is a rationale for civil commitment that does not have to confront the stark contrast between us and them—between committing the mentally ill for their benefit or for ours. One of the ideas I want to advance is that behaviors that we find objectionable can in themselves be evidence of mental illness. Hospitalization in such cases benefits the mentally ill by enabling them to cope with exactly these behaviors.

A Case for Commitment

Consider the following (mostly hypothetical) cases.

Case 1

Arno Blocher is 28 years old and suffers from schizophrenia, although the subtype is unclear. At times he has delusions, and at other times, he seems to be disorganized in speech and behavior. He is clearly better at some times and worse at others. Arno lives with and is well taken care of by his family (i.e., his parents). He is not being treated for his illness—he does not want treatment and his family does not want it for him (largely because he does not want it). His family is willing, as most are not, to put up with whatever problems the mental illness involves.

Arno's mental illness is treatable, at least in the sense that there are well-understood treatments that are available, recommended, and regularly used by the psychiatric profession. These treatments are regarded by (most) psychiatrists as helping people with their delusions, disorganized speech, anxiety, depression, or whatever. However, they do not effect cures. Symptoms are either diminished or eliminated for periods of time, with perhaps recurring episodes. On the other hand, there are often unwelcome side effects, ranging from insomnia to potentially dangerous conditions. Arno and his family do not think the benefits of treatment are worth the downsides: Arno would rather have the illness untreated than a diminished form of the illness along with the side effects, and his family goes along with this. Arno has a job that he can do reasonably well; he works in his father's machine shop. He can function socially to some extent, or at least periodically.

It is important to bear in mind that despite his illness, Arno is competent. Advocates for the mentally ill have worked hard, and successfully, to make the point that mental illness is not the same as incompetence and that mentally ill people can be competent to do many things, including making decisions about their care. Competence is usually explained as comprising a variety of capacities that one must have to some appropriate extent abilities to absorb information, to reason, to evaluate, and to make decisions. Possession of these capacities, however, does not necessarily extend to all areas of one's life, for a person may be competent with regard to certain things but less so with regard to others. Thus, although someone having a mental illness may have diminished competence in some areas of life, it does not follow that there is diminished competence to assess properly the benefits of hospitalization and treatment. But although it does not follow that a mentally ill person lacks the capacity to make a treatment decision, it has been suggested that the degree to which a person must have certain abilities in order to be regarded as competent depends on the *degree of risk* of failing to make the decision to accept hospitalization.[21] Even with this criterion of competence, however, Arno Blocher is competent. Nothing that has been said about him points in the other direction unless, that is, his refusal of hospitalization is taken as conclusive evidence of incompetence. Such a move should of course be resisted (although some critics think it is the normal assumption of the medical profession), for otherwise we would have to say that having a mental illness that the psychiatric profession thinks treatable inevitably compromises the competence to make the health care decision.

Is there a case for involuntary civil commitment and treatment in this situation? Is there a case for overriding the judgment of Arno and of his family? Let us suppose that Arno and his family are mistaken—he would in fact be better off with treatment. Most psychiatrists would think so and believe that there is evidence for it; let us assume they are right. There is, of course, the question of what this means—does it mean that Arno would be better off in that, ultimately, he would concur? Or that he would not but others would judge him to be better off? Let us suppose that it is the former.

If need-for-treatment is a legitimate, and sufficient, ground for commitment, then it must apply not only to the homeless and those whose families cannot or will not care for them, but also to Arno. However, as things presently stand in the psychiatric profession and given what we presently believe about what constitutes a person's good and who is responsible for ascertaining it, I assume there is no case for forced commitment and treatment.

Compare Arno with someone (Camille Derby) suffering from severe obesity who would rather remain obese than undergo what it would take to become nonobese. Camille knows what the deficits of obesity are, both to her health and socially. She knows the way to change, but, having tried it before, she does not think that it is worth it to put herself through what is involved. She may be mistaken about this, but even if she is, I assume there would be no case for forced deobesification. This seems quite obvious, yet, despite the clear parallels (i.e., the deficits associated with the condition and the fact that something can be done about it), many find it less obvious when mental difficulties are involved. This may reflect the belief that people with mental illnesses either do not properly understand the deficits from which they suffer or do not properly understand that their situations can be improved.

As to the first, it is no doubt true that some people, owing to their illness, do not understand that they are ill (psychiatrists call this absence of insight); they might think themselves ordinary, or special, or simply different. However, this is not always the case,[22] and Arno, as described, does understand that he has an illness. As to the second, Arno, as described, does not properly understand that treatment will be a net gain. This, however, is irrelevant to the issue of forced commitment and treatment, for Camille is similarly mistaken and (I am assuming) no one would think of forcing her to lose weight. However mistaken, Camille is competent to make this decision for herself *and so is Arno*.

Let us inquire, then, as to what alterations of Arno's situation would begin to make involuntary commitment and treatment suitable (we will not consider treatment as separable from commitment, as the law presently does).

Case 2

Elvin Fancher has the same mental illness as Arno but does not have the same family support. He is not homeless or poor—he has a house and income from a trust fund—but has no one to care for him, nor does he want anyone to care for him. Elvin keeps to himself and bothers no one; he sits on his porch and smiles at people and sometimes says nonthreatening things to passersby. People think he is crazy. He offends against no social norms.

Case 3

Galen Howard is similar to Elvin Fancher in having the same mental illness as Arno and in not having family support. He, too, is not homeless or poor—he has a house and income from a trust fund—but has no one to care for him and does not want anyone to care for him. Galen prefers to spend most of his time outside his house, on the sidewalk, where he "interacts" with passersby. He is known to a few who live nearby or pass by regularly and has brief exchanges with them, but, for the most part, he approaches or accosts strangers and tries to engage them in conversation. He sits or lies down on the sidewalk, mumbles to himself, and seems to others to be deranged. He is dirty, smelly, offensive, and a bit frightening to some of the people with whom he "interacts."

Case 4

Joyce Brown has already been described: She lived on a hot air grate in an affluent neighborhood on the East Side of Manhattan, "interacting" with people on the street. She was given money, some of which she used to buy food and some of which she tore up. She made obscene gestures to some passersby. At night, she wrapped herself in a sheet and slept on the grate. It was generally presumed that she was mentally ill, although she and her supporters maintained that she was not. There seems to be no doubt that she was intelligent and she had some education; she said she could get a job and a regular place to live if and when she chose. In the meanwhile, she lived on the street. She urinated and defecated on the street and sometimes on herself. She was ragged, dirty, smelly, offensive, and a bit frightening to some of the people with whom she "interacted."

Case 5

Ian Jarman is a lot like Joyce Brown, except that his homelessness is definitely not a matter of choice. Yet, he too rejects treatment.

In which of the foregoing cases, if any, would civil commitment be justified? If we say that a sufficient justification of civil commitment is that it is for the benefit of the mentally ill person, then there is a problem with regard to Case 1, for Arno would benefit from hospitalization and treatment. However, as we have said, Arno is competent to make this decision for himself and is entitled to make the wrong decision about what is in his best interest. (There is no need in this discussion to confront the possibility that Arno rejects treatment but his parents insist on it, or that Arno wants treatment but his parents reject it.) The same line of thought applies to Elvin Fancher: His situation is a bit more precarious than Arno's, but he is competent to make decisions for himself.

Those opposing the commitment of people like Joyce Brown argue that need-for-treatment is not the issue. The real reason such people are committed, it is claimed, is that we do not like their behavior and want to get them off the streets, especially in affluent areas. What is wrong with this, it is argued, is that committing people for this reason is for our benefit, not theirs, and that civil commitment thus looks more like punishment than treatment. This is certainly Szasz's view: He charges that psychiatrics have allowed themselves to become agents of the state for the controlling of undesirable behavior.[23] Szasz believes, instead, that if we do not like certain behaviors, we should openly make them crimes and punish violators as offenders, repeatedly if necessary, but we should not force drugs on them.

Let us use this position as a starting point. Suppose that it was true that Joyce Brown defecated on the street and suppose that this was illegal. A digression might be useful here. Brown's attorneys argued that her defecating on the street was not an indication that she could not care for herself (an aspect of the danger-to-oneself standard for civil commitment). Rather, they said, it was a rational response to the circumstance that there were no public toilets and that owners of stores and restaurants would not allow her to use their facilities. If, however, we ignore the danger-to-oneself issue and assume that the problem with her behavior is that it offends against social standards and is illegal, then the fact that there were no public toilets and that she was denied access to private ones is irrelevant. There may be good reasons for a city to provide public toilets, but no one has a personal right that they be provided. Her need for regular access to private facilities stems from her having no facilities (i.e., a home) of her own. This was represented by Brown and her attorneys as a matter of choice—she could have had a home, but chose to live on the street. Against this, however, I would argue that this is a choice she is not entitled to make. People may reasonably be expected to take care of their own basic needs, including (and especially) their excretory needs. In dealing with these needs, no one is entitled to impose on others, either by excreting publicly or by insisting on using private facilities.

So, if we do not like people defecating on the street, perhaps we should, as Szasz suggests, make it illegal, and then Brown should be arrested and punished, but psychiatry should be left out of it. Suppose that Brown goes to jail and that while in jail she uses the toilet that is provided. This would confirm her attorney's position that she defecated on the street because she could not find a toilet to use. And if, on being released, she continued to defecate on the street, she would continue to be subject to arrest, just as any other offender.

However, suppose instead that the offending behavior is *not* freely chosen—that there is a mental illness and that it compromises the capacity to control certain behavior (which *might* be indicated, for example, by a failure to use a toilet even when there is one available). In some criminal cases, an offender makes this an issue by pleading not guilty by reason of insanity—and when the plea is believed and accepted, the offender is committed for psychiatric treatment. However, there are very many cases in which there is no plea of insanity and yet there is a suspicion of mental illness. Indeed, it is frequently claimed that substantial numbers of people in our jails are mentally ill and that since deinstitutionalization, jails have come to hold more mentally ill people than all other facilities.

This suggests a middle ground between socially centered and paternalistic grounds for civil commitment. The idea is that as much as we might like to help someone who is mentally ill, if the person does not want to be helped and is not violating social norms (in particular, legal prohibitions), we have no justification for interfering. On the other hand, if mental illness leads a person to offend against social (legal) norms, we may have a justification for interfering, not simply with criminal sanctions but also medically (psychiatrically) so long as we can actually do some good for the person. More specifically, commitment will be justified under the following conditions:

1. There is illegal behavior.
2. The behavior is chronic.
3. There is reason to think that the behavior is brought about by mental illness.
4. It is reasonable to believe that the behavior can be altered by well-understood treatments that are generally regarded as helpful to people with such illnesses.

This proposal for justifying civil commitment is socially centered insofar as it is concerned with behavior that we feel we do not have to put up with, and paternalistic insofar as it is designed to help the offender and is restricted to cases in which there is actually something that medicine (psychiatry) can do for the person. The central point is that it calls upon society to define carefully just what it objects to and to make only such behavior the touchstone for intervening in someone's life. Further, it calls on society to treat violations of social norms as criminal matters until it becomes reasonable to believe that mental illness makes the offender unable or unwilling to control the behavior. It is important that this not be presumed, for it is simply not the case that everyone with a mental illness cannot control his or her behavior.

What, then, makes it reasonable (sufficiently so to warrant commitment) to think the offending behavior may be brought about by mental illness (condition 3)? What is being highlighted in this discussion is that there are cases in which the nature of the offending behavior is *itself* evidence of mental illness. Suppose, for example, that Ian Jarman (or Joyce Brown) defecates not only on the street (perhaps claiming to do it because of lack of access to toilets) but also on the floor of his jail cell despite the availability of toilet facilities. This would, I believe, constitute evidence of mental illness (rebuttable evidence, to be sure, for the behavior might, for example, be a deliberate challenge to the authorities). The point is that there are behaviors that are expected of properly functioning people, that the absence of these behaviors calls for an explanation indicating that, however unacceptable, they are sufficiently within the control of the person doing them that we can believe that his or her functioning is not impaired, and that in the absence of a convincing explanation it is reasonable to think that the person is impaired in some way. Offenses like stealing do not ordinarily put into question one's capacity to function adequately, but offenses like public defecation do put this into question. So too would such things as public nudity and screaming at strangers. It is worth noting that treating such behaviors as evidence of mental illness means that there may be a social element in mental illness, for different societies may regard different behavior as indicative of failure to be a properly functioning person. This feature of the notion of mental illness only intensifies the need to be cautious with the idea of mental illness, which risks being abused politically. Nevertheless, certain (but certainly not all) failures of socialization can legitimately be regarded as aspects or evidences of mental illness.

How does the proposal apply to such problems as use of illegal drugs, sexual offenses,[24] and even writing bad checks, any of which, in at least some cases, may involve mental illness. Condition (4) speaks to such cases, limiting the use of psychiatric intervention and thus the possibility of abuse. The mere fact that psychiatric treatment (i.e., drugs) can prevent offending behavior is not enough; people in a stupor might not do these things, but merely stopping them in this way is not helping them. If we can not *help* them medically, we should, as Szasz insists, limit ourselves to punishing them when they commit offenses. This is the paternalistic side of the basis for commitment.

It may be worth noting that the proposal does not speak to the question of competence. Requiring incompetence for civil commitment would be relevant on a danger-to-self criterion for commitment. However, here the basis for commitment is not danger to oneself, but offending (lawbreaking) behavior under certain conditions.

Finally, how would the proposal apply to the five cases outlined earlier above? In Cases 1 and 2 (Arno Blocher and Elvin Fancher), civil commitment is not warranted, for there is no criminal offense. We may not deprive of their liberty people who do not do things we are explicitly willing to prohibit. In Cases 4 and 5 (Joyce Brown and Ian Jarman), commitment is warranted only if some of their behavior is illegal and they have been punished for it but still continue to do it (so long as the behavior is reasonably regarded as uncontrollable and not, for example, a deliberate political protest). Case 3 (Galen Howard), as described, is similar to Cases 1 and 2. What it highlights is that if behavior is regarded as unacceptable and the person engaging in the behavior is regarded as mentally ill, we cannot legitimately do anything about it if we are not willing to prohibit the unwanted behavior. Everyone, mentally ill or not, must first know what is expected of them and have the opportunity to conform to the requirements. Decades of experience with mental illness have shown that the existence of mental illness does not automatically make a person unable to adhere to requirements.

Conclusion

Dangerousness is too limiting a criterion for civil commitment. There are behaviors that the public does not have to tolerate and that can legitimately be made illegal, even though they are not dangerous. How should society deal with persistent violations of such prohibitions? One way is through the criminal law—apprehend violators, try them with the full panoply of legal procedure, and punish those found guilty, repeatedly if necessary. At some point, though, it becomes reasonable to wonder whether there is a mental disorder underlying or associated with the behavior. Even without a plea of "not guilty by reason of insanity," it sometimes becomes reasonable to initiate civil commitment proceedings to determine whether the objectionable and illegal behavior is the result of a mental illness and whether the medical profession can help the person avoid such behavior. The penal system is not designed for such individuals; it presupposes offenders who can control their behavior. A person who offends against the law should certainly, in the first (and perhaps the second and third) instance, be dealt with as someone who is responsible and can control his or her behavior. Repeated offending behavior, however, and especially behavior of certain sorts, begins to suggest that it is beyond the purview of the penal system. This is not, however, a recipe for moving straightaway to the psychiatric control of offenders. In cases of this sort, all of the recently developed civil liberties protections (criminal-style procedure) need to be respected and the focus must be on the health of the person being dealt with rather than on the offense, even though it is the offense that has given rise to the suspicion that mental illness lies behind the offending behavior.[25]

Notes and References

[1]Miller, G. E. and Iscoe, I. (1990) A state mental health commissioner and the politics of mental illness (Hargrove, E. C. and Glidewell, J. C., eds.), *Impossible Jobs in Public Management,* University Press of Kansas, Lawrence, p. 104.

[2]Campbell, T. and Heginbotham, C. (1990) *Mental Illness: Prejudice, Discrimination and the Law,* Dartmouth, Aldershot, UK, p. 21.

[3] Szasz, T. S. Statement and Manifesto, http://www.enabling.org/ia/szasz/manifesto.html. March 1998.

[4]Szasz, T. S. (1998) *Cruel Compassion,* Syracuse University Press, Syracuse, p. 91.

[5]Tucker, W. (1990) *The Excluded Americans,* Regnery Gateway, Washington, DC, p. 69.

[6]Consider, for example, pathological gambling, mathematics disorder, caffeine-induced sleep disorder, adjustment disorder, and nicotine-related disorder. Most of these are not mentioned in *A Guide to Treatments That Work* (1998), Nathan, P. E. and Gorman, J. M., eds., Oxford University Press, New York.

[7]Curing the therapeutic state (interview with T. S. Szasz), *Reason*, July 2000, p. 31.

[8]"I . . . *define illness as the pathologist defines it—as a structural or functional abnormality of cells, tissues, organs, or bodies.*" Szasz, T. S. (1987) *Insanity: The Idea and its Consequences,* J. Wiley, New York, p. 12.

[9]Szasz, T. S. *Cruel Compassion,* p. 162.

[10]*Lessard v. Schmidt*, 349 F. Supp. 1078 (E. D. Wis. 1972).

[11]Appelbaum, P. S. (1994) *Almost a Revolution,* Oxford University Press, New York, p. 28. However, note that Appelbaum maintains that the change from "need-for-treatment" to "dangerousness" has not actually resulted in fewer commitments. *See* pp. 33–57.

[12]*Wyatt v. Stickney*, 344 F.Supp. 387 (M. D. Ala. 1972), affirmed in part 503 F.2d 1305 (5th Cir., 1974).

[13]*O'Connor v. Donaldson*, 422 U. S. 563, 95 S.Ct. 2486 (1975).

[14]It is worth noting that part of what drove the process of deinstitutionalization was the issue of who would be paying. Costs of running the large mental hospitals were borne by the states and were often a very large part of the state budget. The Community

Mental Health Program, by contrast, was a federal program, paid for out of Social Security funds.

[15]Attributed to Darryl Treffert. *See* Appelbaum, P. S., p. 30.

[16]Appelbaum, P. S., p. 32.

[17]Quoted in Pence, G. E. (2000) *Classic Cases in Medical Ethics,* 3rd ed., McGraw-Hill, Boston, pp. 375–376.

[18]*See* Pence, G. E., pp. 368ff.

[19]Some say, though, that at the present time the issue is moot owing to a failure of resources: "The reality is that commitment is no longer much of a civil liberties threat—state hospitals don't want patients, and the short-term hospitals can't get rid of them fast enough. If New York had outpatient commitment, it probably would not be enforced because the resources aren't there; or if it were enforced, given current resources, one group of people that needs care would be pushed aside for another group that also needs it. There are 37 states with outpatient commitment laws, and most have systems as dysfunctional as New York's." Winerip, M. (1999) Bedlam on the streets. *New York Times Mag:* May 23, p. 70.

[20]Szasz, T. S. *Cruel Compassion*, pp. 199–200.

[21]Buchanan, A. E. and Brock, D. W. (1990) *Deciding for Others: The Ethics of Surrogate Decision Making,* Cambridge University Press, New York, pp. 317ff. For a different view, *see* White, B.C. (1994) *Competence to Consent,* Georgetown University Press, Washington, DC, p. 113.

[22]Even when there is lack of insight, it is questionable whether it is appropriate to use this as part of the basis for overriding the wishes of a mentally ill person. *See* Olsen, D. P. (1998) Toward an ethical standard for coerced mental health treatment: least restrictive or most therapeutic? *J. Clin. Studies* **9(3)**,235–246.

[23]*See* Szasz, T.S. *Cruel Compassion*, Chap. 11.

[24]The recent case of *Kansas v. Hendricks*, 117 S.Ct. 2072 (1997), involving the civil commitment of persons who are likely to engage in "predatory acts of sexual violence" owing to a "mental abnormality" or "personality disorder" can be seen in light of the proposed criterion, although the court's due process discussion presumes the established dangerousness standard.

[25]I would like to thank Professor George Graham for helpful comments.

Abstract

Thirty-seven states have passed legislation establishing involuntary outpatient commitment programs. These programs seek to identify mentally ill people who are at risk of becoming violent, devise programs of treatment for them, and authorize the use of law enforcement personnel to ensure that they hew to the treatment programs. Those singled out for treatment would remain in the community but would be enrolled in treatment programs whether or not they gave consent.

These programs are at the center of intense debate focused on four issues—those of whether coerced treatment programs are effective, whether the mentally ill are more violent than the general population, whether mental health professionals have the means to identify patients who are at risk for violent behavior, and whether states are morally justified in overriding an individual's right of consent to treatment in order to enhance public safety.

An examination of recent studies reveals no evidence that involuntary outpatient treatment programs are less effective than voluntary programs. A growing body of research shows that mentally ill persons whose symptoms are active are more violent than the population at large and that mentally ill persons who also fall prey to substance abuse are considerably more apt to be violent than are other mentally ill people or healthy persons. Mental health professionals have reasonably effective means to identify groups of mentally ill people who are likely to become violent, and they are steadily working to improve the accuracy and efficiency of these measures.

However, overriding mentally ill people's right to consent to treatment in hopes of enhancing public safety is unjustified because it requires the mentally ill to carry a far greater burden than other members of society are willing to bear. Mentally ill people are responsible for less than 1000 homicides in the United States each year, but over 40,000 Americans died in automobile accidents and over 30,000 died from injuries caused by firearms. Nonetheless, the public is unwilling to accept legislation restricting automobile travel or firearm ownership which could easily save far more lives than involuntary outpatient commitment programs and would not violate any fundamental legal or moral rights. Hence, involuntary outpatient commitment programs are morally unjustified because they are inequitable.

Involuntary Outpatient Commitment

Gerard Elfstrom

Shortly after 5:00 PM on January 3, 1999, Kendra Webdale was waiting at the northbound subway stop at 23rd Street and Broadway in New York City. She was approached by a disheveled young man who apparently asked her the time. She turned away from him as the N train pulled into the station. At that instant, the stranger shoved her off the platform and onto the tracks. She was killed instantly.[1] Webdale was the sort of person who elicits intense popular sympathy. Blond haired, blue-eyed, youthful, good-natured, and outgoing, she had a diverse array of devoted friends. Her assailant, Andrew Goldstein, was a pudgy, unkempt loner who had few acquaintances and was unable to hold a job. Worse, he was mentally ill, having suffering from schizophrenia since he was 16 years old. He was first admitted to inpatient treatment at Creedmoor Psychiatric Center in 1989 and was bounced in and out of a variety of institutions and treatment programs until mid-December 1998. By the time he pushed Webdale off the subway platform, his medical files were 3500 pages long.[2] He also had a well-established history of violent assaults. Most involved only kicking, punching, and shoving, but several of his

From: *Biomedical Ethics Reviews: Mental Illness and Public Health Care*
Edited by: J. Humber & R. Almeder © Humana Press Inc., Totowa, NJ

attacks sent their victims to the hospital.[3] Goldstein gave much the same explanation after each incident of violence, namely that he was inhabited by a superior force that, against his will, forced the acts.[4] After the assault on Webdale, he used similar language in his confession. He said, "I felt a sensation, like something was entering me. . . . I got the urge to push, shove or sidekick. As the train was coming, the feeling disappeared and came back. . . . I pushed the woman who had blonde hair."[5]

Webdale's death sparked an enormous public outcry in New York City and across the nation. The public was stunned because Webdale was an attractive person, the incident seemed absolutely senseless, and it stoked anxiety about the presence of the mentally ill in local communities. The stir intensified when the public became aware of Goldstein's long history of mental illness and violent assaults. Outrage of this magnitude energizes politicians and prompts them to loosen the public purse strings. So, New York's Governor Pataki and the State Legislature busied themselves with several measures. One pledged an additional $125 million to expand the system of community-based services for the mentally ill, construct housing for them, and reverse New York State's policy of gradually shutting down its inpatient hospitals. Mental health care advocates and professionals applauded these measures, but they were far less enthusiastic about a second piece of legislation, known as Kendra's Law. Kendra's Law established machinery to identify mentally ill persons in danger of becoming violent, devise a plan of treatment for them, and authorize use of law enforcement personnel to ensure that they hew to the prescribed regimen whether or not they consent to it.[6] For several reasons, the law does not require that they be placed in an institution. Rather, it allows them to remain in their communities. First, the individuals selected for supervision under this program pose no immediate threat of violence. Second, mental health professionals believe that many benefit from living as near a normal life as is possible. Finally, the mentally ill commonly prefer living independently. The process established by Kendra's

Law is referred to as "involuntary outpatient commitment." New York is not alone in finding this idea attractive. Some 37 states also have involuntary outpatient commitment laws, although their details commonly differ from New York's legislation.[7]

Kendra's Law ignited broad and spirited controversy that focused on several issues posed by involuntary outpatient commitment. First, partisans offer several arguments to support the claim that involuntary outpatient commitment is futile, that is, it is ineffectual. Second, many advocates assert that the effort is misguided because the mentally ill are no more dangerous than the population at large. Third, many claim that it is inept because mental health professionals have consistently proven they are unable to predict which of their patients are likely to become violent. Last, critics assert that involuntary outpatient commitment is morally unjustified because it overrides a fundamental moral and legal right, that of consent to treatment, and it allows the state's instruments of coercion to be deployed against persons who pose no threat of immediate harm. The first three issues concern matters of fact. Only the last introduces questions of morality and public policy. However, the first three issues address matters that are vitally important to examining questions of policy and moral justification. Hence, each will be examined in turn.

Matters of Fact

Futile

Those convinced that outpatient commitment legislation is futile cite three arguments to support their claim. First, they point out that outpatient commitment machinery would have been completely unnecessary in Goldstein's case. He had not refused treatment. Instead, he sought it. In the two years prior to his assault on Webdale, he received inpatient treatment on 13 occasions. He voluntarily entered treatment on all 13 admissions.[8] Critics of outpatient commitment appear to infer from this that all of the mentally ill who are potentially violent would voluntarily accept treatment if it were available to them. However, although it seems perfectly reasonable to assume that many of them would, it seems unlikely that *all* would do so. In fact, several mental health researchers assert that the most dangerous patients with the most pronounced history of violence are also those most likely to resist treatment.[9]

Second, some commentators assert that Kendra's Law legislation is futile because coerced treatment will prove ineffectual. They state that successful treatment requires the active and voluntary participation of the patient and a relationship of trust and mutual respect between the patient and mental health professionals. However, if patients are coerced into treatment programs, they will not trust their counselors and will not actively participate in treatment plans.[10] These are important and plausible claims. However, although mental health researchers, patient advocates, and policy experts are engaged in a furious debate on the question of whether patients assigned to involuntary outpatient commitment fare better than those who voluntarily enroll in outpatient treatment programs, no empirical data support the conclusion that those assigned to involuntary outpatient commitment programs fare worse than those who enroll voluntarily.[11] One administrator has also reported that those selected for outpatient commitment reported feeling no more coerced than members of control groups who voluntarily participated. The key factors determining whether patients felt coerced were those of whether they believed mental health professionals dealt with them honestly and fairly.[12] Hence, the empirical data do not support the claim that outpatient commitment will necessarily be ineffectual.

Finally, at least one knowledgeable and shrewd observer believes that outpatient commitment legislation will prove ineffectual simply because governments are unlikely to provide sufficient funding to establish these programs and are unlikely to sustain funding for them once moments of crisis slip from public attention.[13] This is an important and astute observation. However, it is not plausible to presume that such programs will never receive adequate funding. Further, even if they do not, the scraps and rags of programs in place, such as those in New York State, will continue to function. To the extent that they do, individuals will be pressed into outpatient commitment and the issues generated by Kendra's Law will remain in contention.

Misguided

For many years, patient advocacy groups and mental health professionals have asserted that the mentally ill are no more violent than the population as a whole. If this is the case, legislation such as Kendra's Law is both unnecessary and pernicious. If the mentally ill are no more violent than anyone else, it is unnecessary, because there is no reason for singling them out for special attention on grounds of their violent nature. It is also pernicious because it reinforces the stigma of violence that the mentally ill carry. Mental health advocates assert that those who are mentally ill carry a large stigma simply by virtue of their disease. If they must shoulder the additional stigma of being labeled violent, their lives will become more difficult and the obstacles in the way of their recovery will increase.

In recent years, however, evidence has accumulated that supports the conclusion that the mentally ill as a group are indeed more prone to violence than the general population. One commentator notes, "In the last decade, however, the evidence showing a link between violence, crime, and mental illness has mounted. It cannot be dismissed; it should not be ignored," and the *Harvard Mental Health Letter* reports,

> "People with any history of psychiatric treatment are two to three times more violent than average. . . . In a study of 18,000 people conducted by the National Institute of Mental Health, more than half of those who reported committing acts of violence in the previous year, compared with 20% of the general population, met the criteria for a psychiatric disorder in the American Psychiatric Association's current diagnostic manual."[14]

The evidence emerges from several major studies. Studies of birth cohort groups (i.e., groups of people born in the same year) in three Scandinavian countries yielded strong correlations between mental illness and violence. In a Swedish study, males with a history of mental illness were 2.5 times more likely than other males in their cohort group to be convicted of crimes and 4 times more likely to be convicted of violent acts. A Danish study revealed that men with a history of mental illness were 2.4–4.5 times more likely to commit violent crimes than men without a history of mental illness, and those with a history of mental illness along with a history of alcohol abuse were 4.2–6.7 times more likely to commit violent crimes than other males. A Finnish study yielded similar results. Schizophrenics were found to be 3.6 times more likely to commit violent crimes than healthy males, whereas those who exhibited other psychoses were 7.7 times more likely to commit violent crimes than others. Mentally ill people who also engaged in alcohol abuse were 25.2 times more likely to commit violent crimes than healthy males. An elaborate and lavishly funded US study also revealed that recently discharged mental patients were significantly more apt to be violent than the healthy populations in the neighborhoods where they lived.[15] Data also show that mental health care workers in the United Kingdom are 3 times more likely to suffer violent assault while on the job than industrial workers and that 1 in 10 hospitalized mental patients commit violent acts.[16]

Nonetheless, the picture revealed by the above studies is complex and nuanced. The mentally ill whose symptoms are not active are no more apt to be violent than the population as a whole. However, those who fail to take their medications and whose symptoms become active as a result are more likely to be violent than healthy persons. Worse, mentally ill people whose symptoms are active and who fall prey to substance abuse are significantly more prone to violence than those whose symptoms are active but who do not engage in substance abuse. Further, the mentally ill are more likely than the public at large to fall prey to alcohol or drug abuse and are more likely to become violent when they do so.[17] Also, the mentally ill often dislike their medications, and, hence, they do not take them. Andrew Goldstein is a good example. Although he had no record of drug or alcohol abuse, he was notoriously lax in maintaining his schedule of medications. He, as do many other patients, reported that the medications made him listless, caused soreness, prevented him from sleeping, and made his mouth dry. So, left to his own devices, he did not take them. When he lapsed, the symptoms of his disease reappeared and the episodes of violent behavior also recurred.[18]

As a result of these findings and likely also as a result of the popular belief that something must be done to address violence on the part of the mentally ill, several mental health professionals have determined that it is best to acknowledge the problem and seek to address it honestly and effectively.[19] Paradoxically perhaps, some prominent researchers believe several patient advocates now also embrace the idea, apparently on grounds that they believe the general public is more apt to support increased funding for mental health services if it is persuaded that doing so will reduce violence by the mentally ill.[20]

Inept

Historically, mental health professionals have vigorously asserted that they have little competence to determine which individuals are likely to cause harm and which not. Since 1984, the American Psychiatric Association has stated as policy that the profession of psychiatry has no expertise that enables it to identify which patients are prone to violence.[21] Further, it has often been claimed that mental health professional's estimates of the danger of violence posed by mentally ill individuals are no more accurate than random guesses.[22] A variety of recent studies yield data that undermine this assertion. The studies do point out that it is difficult to gain solid information about success in predicting when particular patients are likely to become violent.[23] Partly, this is because professionals who have reason to believe a patient will become violent will likely take measures to prevent an outbreak. Furthermore, some studies have shown that judgments about violence are guided not simply by estimates of probability but also by underlying principles of judgment that will vary from practitioner to practitioner. Some, for example, believe that it is best to err on the side of caution in seeking to prevent violent acts by the mentally ill, whereas others are convinced that it is enormously important to allow patients maximal freedom.[24] Despite the difficulties, a number of elaborate research programs have undertaken to identify the factors most closely associated with violent behavior by the mentally ill and to radically increase the accuracy of attempts to predict violent behavior. One study was able to sort some patients into a group that was half as likely to be violent as the group of patients as a whole, and it was able to sort other patients into a group that was twice as likely to be violent as the entire population of patients. They reveal that patients who have a history of engaging in violent acts and have fantasies about violence have a high risk of engaging in violent behavior, whereas those lacking these characteristics are at low risk of engaging in violent behavior.[25] Several studies have claimed considerable success in this endeavor, and researchers are working to refine decision protocols in ways that will allow mental health practitioners to quickly and easily determine which of their patients are prone to violent behavior. Further, as reported earlier, patients outside institutions who fail to take their medications and who engage in substance abuse are quite likely to become violent. Of course, none of the studies is able to predict which particular patients will become violent. They can only sort patients into groups of those prone to violence and those unlikely to become violent.

Also, the social and legal contexts in which mental health professionals operate has changed over the years. The landmark Tarasoff Case in California of 1976 gave firm legal foundation to the principle that mental health professionals are legally accountable to third parties likely to be harmed by their patients.[26] The principle of patient confidentiality is overridden in such cases. Subsequent legal decisions and the public's outrage over spectacular acts of violence by the mentally ill have strengthened and broadened the responsibility of mental health professionals to be alert to indications that their patients may harm others and take measures to protect those who are at risk of becoming victims.[27] Given continued public interest in this issue, the array of funding available for research on the topic, and their own self-interest, it is likely that mental health professionals will continue to improve and refine their instruments for predicting violent behavior by their patients.

Rights

The above survey of questions of matters of fact reveals that involuntary outpatient commitment is not futile, the mentally ill genuinely are more dangerous than the population as a whole when they fail to take their medications and when they engage in substance abuse, and mental health professionals have far greater resources for identifying those groups of the mentally ill who are likely to become violent than has been previously recognized. Hence, involuntary outpatient commitment legislation cannot be ruled unworkable simply by appeal to matters of fact. Nonetheless, the moral and legal issues remain, and they remain formidable. Involuntary outpatient commitment legislation establishes machinery that allows government officials to override the right of consent to treatment of a group of people who have not been deemed incompetent to consent to treatment and who do not pose an immediate threat of harm to others. Further, the legislation makes provision for employing the state's instruments of coercion to enforce these decisions. The justification offered for violating the rights of the mentally ill in this manner is that public safety will be enhanced, because it is hoped that fewer people will be killed or injured by the mentally ill as a result of these measures.

Public Safety vs the Right of Consent

Balancing fundamental human rights against concern for public safety is a difficult matter. Few are prepared to assert that concern for human rights should completely set aside concerns for public safety. "The Constitution is not a suicide pact," is a commonplace of legal analysis.[28] Even ardent advocates of the rights of the mentally ill agree that the mentally ill who are deemed to pose an immediate danger to others may have their right of consent to treatment overridden.[29] Further, the concern for public safety is grounded on recognition of fundamental human rights—perhaps the most fundamental of all—those of the right to life and security. Nonetheless, few in the United States are prepared to sacrifice all other human rights on the altar of public safety. There are many ways to enhance public safety at the cost of overriding human rights, such as trimming rights to due process, privacy, freedom of speech, and freedom of association. Few are likely to agree that public safety concerns are sufficient to outweigh any significant portion of these rights. Furthermore, it is beyond controversy that the right of consent to treatment is one of the fundamental legal and moral rights of citizens of the United States. So, concern for public safety is balanced against other human rights. The difficulty is to strike the balance between the two groups in reasonable and morally justifiable fashion. Several considerations may assist deliberation about whether Kendra's Law and other legislation like it strike a defensible balance between the right to consent to treatment and the requirements of public safety.

The first place to look for a clue is the circumstance where all agree that the right of consent to treatment may be overridden on grounds of public safety; this is when the mentally ill pose an immediate threat of harm to others. If mental health professionals have the resources to determine which individuals are likely to do harm to others, there seems little reason to defer action until a crisis is at hand. No prudent professional group or office of policy-makers waits until crisis erupts to take preventative measures. They seek ways to anticipate threats to public safety, and they devise plans to forestall disaster. Those who insist that the right to consent to treatment can never be overridden for those who are not an immediate threat to others put public officials in a difficult position. They may be able to identify individuals who are highly likely to become an immediate danger to others if they do not take their medications and may be well aware that these individuals have a record of failure to take their medications, but they cannot take measures to assure that these individuals take their medications. Hence, strictly hewing to the policy of honoring the right to consent to treatment for those not an immediate threat to others forces officials to wait until the crisis erupts and lives are endangered before they take coercive action.

So, if the mentally ill identified as potentially dangerous are relevantly like those who are an immediate danger to others, there is little justification for honoring the right to consent to treatment for the former group but not the latter. However, there are three vitally important differences between the two groups. First, those who pose an immediate threat to others have created an emergency situation, and it is a commonplace that many important rights and restrictions can be set aside in emergencies, as the US Supreme Court's "clear and present danger" test illustrates. Vitally important constitutional rights, such as those of freedom of speech, can be set aside in emergencies, but not otherwise. Those who only pose the danger of becoming violent have not created the emergency conditions that would justify setting aside fundamental rights. More importantly, protecting the right to consent to treatment does not imply that public officials are barred from taking any measures at all to prevent violence. Certainly formulating plans to identify those most at risk of becoming violent and working to get them enrolled in programs of effective treatment are both prudent and justified. However, overriding important individual rights in a nonemergency situation is unjustified. Second, when individuals pose an immediate threat of harm, officials have no difficulty knowing with precision that these individuals are dangerous. Mental health professionals are working to improve their techniques for identifying individuals at risk of becoming violent, and they have achieved considerable success in doing so. The difficulty is that even the best of these measures can identify only those most likely to become dangerous. No matter how refined and sophisticated the techniques become, they will never be able to identify individuals who certainly will become dangerous. Furthermore, there is no legally or morally justifiable means of establishing the accuracy of these techniques, as responsible authorities will take measures to prevent the condition of those identified as potentially violent from deteriorating to the point where they pose an immediate threat of danger to others. United States law does not allow individuals to be punished by law or forfeit fundamental rights simply because they are at risk of violating the law or causing harm to others. There is no justification for setting aside this principle for a group of people who differ from the general population only by being mentally ill. Finally, when individuals pose an immediate risk of harm, their likely victims can also be identified with precision and measures can be taken to protect them. When the risk of violence is only potential, this effort is vastly more complicated. Therefore, those mentally ill who pose an immediate threat of harm are not relevantly like those who are only identified as being potentially violent, and there is no justification for overriding the right of consent for both groups. Because the mentally ill who pose an immediate threat of harm to others are relevantly different from those who do not, public officials and public policy are not justified in treating the two groups alike.

One issue that has thus far been overlooked is that of the degree of threat to public safety the mentally ill pose. The violence that most completely captures the public's attention is homicide, as the Goldstein case illustrates. Patient advocates for the mentally ill have been quick to cite data. An average of 19,431 homicides were committed in the United States in the 10 years from 1989 to 1998. Of that number, it is estimated by one authority that less than 1000 are committed by people who are mentally ill, a small fraction of the total.[30] Hence, even if the entire group of homicides committed by the mentally ill were eliminated, the number of homicides committed in the United States each year would decrease very little. The United States, in other words, would not be made significantly safer even if all the Andrew Goldsteins and potential Andrew Goldsteins were removed from the population or treated successfully. The difficulty is that violence by the mentally ill is like airplane crashes. Both capture the public's attention and stir its anxieties, even though airplane crashes account for a miniscule portion of the accidental deaths in the United States each year, and homicides by the mentally ill account for a small fraction of the homicides in the United States each year.[31] They are alike because a public response of this magnitude makes these cases important. Because they disturb the public, government officials must take measures to address these deaths. The question is whether the public outrage has sufficient moral weight that public officials are justified in endorsing the violation of individual rights that occurs when an individual's right of consent is overridden and individuals are subjected to coercion by state agencies. The obvious answer to this question would appear to be "No." Public outrage creates practical problems for public officials, but the mere fact of outrage has no moral weight. In fact, among the most basic principles of government in the United States is that those persons whose moral and legal rights are vulnerable need to be protected against popular dislike and suspicion.

If the degree of public outrage is not a reliable instrument for establishing a morally justifiable balance between public safety and human rights, the answer must be sought elsewhere. Considerations of equity may serve the purpose. Any morally justified balance between public safety and human rights must be equitable. An equitable balance would not unjustly harm one group of people, and it would not impose greater burdens on one group of people than all the members of society would be willing to shoulder. This approach derives from a fundamental principle of moral philosophy, namely that all person's interests have equal weight and no one's interests should be cast aside in favor of the interests of the majority. So, to address the question of how to balance the concern for public safety against the concern to protect the right to consent to treatment, the search for an equitable policy must discover what rights the public at large has been willing to yield or to what degree it has proven willing to sacrifice them in order to protect public safety. Firearms and automobile travel serve as useful test cases. Automobile accidents claim far more lives and cause vastly more injury than do mentally ill persons. Every person in the United States is at far greater risk of being killed or injured by an automobile than harmed by someone who is mentally ill. According to the US Centers for Communicable Disease, 43,501 Americans died in automobile accidents in 1998, and 4,277,000 people in 1997 received injuries in automobile accidents that were sufficiently serious to require visits to hospital emergency rooms.[32] Hence, the threat to public safety and the welfare of each individual is far greater than the threat posed by the mentally ill. However, Americans resist paying significant amounts of money to make automobiles and highways safer. They have proven willing to accept legislation requiring them to wear seat belts and pay modest sums of money to make automobiles safer. However, they have certainly not been willing to accept the sacrifice of any right that comes near the importance of the right to consent to treatment. The case of firearms is also illuminating. Firearms are responsible for vastly more death and injury each year than are the mentally ill. The Centers for Communicable Disease states that 30,708 US citizens died as a result of injuries from firearms in 1988.[33] Each individual American faces a far greater risk of being harmed by a firearm than someone who is mentally ill. Further, few Americans can claim they need firearms to meet their life's needs. Yet, American citizens have proven unwilling to accept more than modest restrictions on the purchase and possession of firearms.

Hence, in the cases of automobiles and firearms, Americans have proven unwilling either to accept any significant restriction on their rights or pay any great cost either in money or inconvenience to gain increased safety for themselves. This is despite the fact that firearms and automobiles cause far more harm to them than do the mentally ill. Hence, because they have proven unwilling to accept far more modest restrictions on their lives to enhance public safety in the cases of automobiles and firearms, Americans would be unjustified to ask the mentally ill to sacrifice their right of consent to treatment for the sake of increased public safety. Overriding the right of the mentally ill to consent to treatment in order to enhance public safety is therefore unjustified on grounds that it violates equity.

The above does not imply that American citizens and mental health professionals either can or should be sanguine about the dangers to public safety posed by the mentally ill. Although 1000 deaths each year is a small fraction of the tally of fatalities, it is a significant number of lives. Further, it is roughly the same as the average number of number of lives lost in airplane crashes each year.[34] Yet, following an airplane crash, governmental authorities undertake an exhaustive investigation of the incident. Often these investigations cost millions of dollars. Once the investigation is complete, officials commonly make recommendations designed to prevent similar accidents in the future. These recommendations commonly stipulate changes in equipment on airplanes or air traffic control units, modified flight procedures, or alterations in pilot training. Such measures are often costly and impose considerable burdens on pilots, the airline industry, or the Federal Aviation Agency. In the case of the mentally ill as in the case of airplanes, Americans can enhance public safety without violating anyone's rights simply by spending money. Thus far, once the public outcry over a particular incident has died down, they have proven unwilling to do so.

However, if public safety is genuinely as important to the population at large as is claimed, it should be willing to increase the funding. The reason is simple and straightforward. A variety of informed commentators on the Andrew Goldstein case agree that a program of assertive outreach would have sufficed to motivate him to continue his medications and help move him into a productive life—and thus would have prevented his attack on Kendra Webdale. Under programs of assertive outreach, patients live in the community but are assigned a case worker who checks in on them at least once a day, is available to the patient 24 hours a day, and is sufficiently well acquainted with the patient to be able to detect subtle changes in behavior or circumstances of living that indicate that the patient's condition has changed or that the patient has failed to maintain his or her schedule of medication. These case managers typically have no more than 10–12 patients to oversee, so they are able to monitor their patients carefully and gauge their progress. Although the cost is significant, it is less than the cost of inpatient treatment in an institution or of imprisonment—which is where many mentally ill end up. Further, experts in such matters assert that these programs can be sufficiently well designed that they will be attractive to patients. In other words, patients will want them.[35]

It is reasonable to believe that many of the mentally ill at risk of violent behavior would consent to participate in such programs if such were made available to them. Certainly, there is ample reason to believe Andrew Goldstein would have been a willing participant. Hence, there is good reason to believe that a large portion of the harm caused by the mentally ill could be avoided without violating their right of consent to treatment and there are substantial arguments to support the claim that the public is obligated to do so.

The Recalcitrant

An important difficulty remains. As mentioned earlier, it is unreasonable to expect that all of the mentally ill who are potentially violent will consent to treatment, even when serious attempts are made to make the programs of treatment attractive to them. Worse, those who are most prone to violence are also least likely to accept treatment. Thus, it is likely that some fraction of the group of mentally ill who are competent to make choices about their lives but at risk of becoming violent will refuse treatment and eventually cause harm to innocent people. Note that this group will be comprised of people who are competent and are not presently dangerous, but have qualities which make them prone to violence. However, a number of identifiable groups of non-mentally-ill people also share these characteristics, including young males, the poor, and those who have suffered abuse as children. These latter groups nonetheless enjoy the same rights as the population at large, and the institutions of law and government do not allow their rights to be overridden. Hence, because there are no relevant differences between the groups, there is no justification for treating them differently—at least with regard to overriding their rights. Further, because those who refuse an offer of treatment will be only a subset of those mentally ill who are prone to violence, the threat they pose to public safety will be small. So, once again, because the public at large is unwilling to accept significant restrictions on its rights in return for enhanced public safety, it is unjust to demand that this group suffer impingement on its rights. Overriding their right to consent to treatment is therefore unjustified even for this group.

This does not imply that public officials have no reasonable response to make to this group. There is useful precedent in law governing behavior while intoxicated. Under British–American common law, voluntary intoxication does not absolve individuals from guilt for actions committed under the influence. Although some jurisdictions do not allow any appeal to voluntary intoxication as a defense, most allow defendants to present evidence concerning voluntary intoxication to mitigate guilt. A defendant may appeal to voluntary intoxication to address issues of intent, although an appeal of this sort would not suffice to absolve the defendant of guilt altogether.[36] Similarly, the law could stipulate that those mentally ill who are deemed prone to violence but refuse an offer of therapy would be held legally accountable for any acts they performed as a result of their disease. For them, the insanity defense would not be available, although they might introduce evidence of their disease to address issues of intent. It is reasonable to believe that this possibility would suffice to change the mind of some of the competent but reluctant mentally ill who are deemed prone to violence and prompt them to seek therapy after all. The remainder will need to understand that they are to be held responsible for their choice and its consequences. Public officials are not justified in imposing any further restrictions on them, no more than they would for the population of those who are not mentally ill but are prone to violence.

The Claims of Public Safety

The above does not imply the claims of public safety, and potential victims make no claims on the mentally ill or the mental health practitioners who treat them. Programs that are both morally justified and meet the requirements of public safety have several characteristics: They are designed to vigorously search out those patients who are prone to violent acts, make competent programs of treatment available to them, and provide means to direct them into these programs. Mental health professionals are obliged to sensitize themselves to the potential for violence in their patients, continue to work to improve their ability to read portents of violent behavior, and devise more effective programs of treatment for them. Governmental authorities, for their part, are obliged to insist that competent programs of treatment be established for such patients and, more to the point, provide the funding necessary for these programs to function effectively. Further, they are obliged to alter public laws regarding the criminal insanity defense to strike it as an option for those mentally ill who are offered the option of treatment but refuse. The violence-prone mentally ill who refuse treatment are properly in the same category as the voluntarily intoxicated, that is, they have freely chosen their circumstances and shall be legally liable for their acts. If they are competent to give or withhold consent to treatment, they should also be competent to grasp the implications of this additional commitment. Further, this revision of law would introduce in the legal context an element that mental health advocates insist be present in the treatment of the mentally ill, that is, a vitally important element of responsibility for their own fates and for the consequences of their actions.

Objections

These proposals are likely to be met by several objections, all of which deserve comment. The first is a response that mental health advocates have made to schemes for involuntary outpatient commitment. It is that they are unnecessary because the law and mental health care already contain provisions for emergency commitment, conditional release from commitment, and directed living in the community.[37] These options are all important and useful. However, they fail to address the claims of public safety. If maximal public safety is to be sought that is compatible with carefully respecting the rights of the mentally ill, a concerted and organized effort must be made to identify and seek to provide appropriate treatment to those at risk of becoming violent. Further, such programs provide an important benefit for the mentally ill and for mental health professionals. Both groups have expressed concern that the mentally ill be stigmatized as being violent and therefore suffer a considerable burden in addition to their disease.[38] In the eyes of the public at large, the failures of the mental health care system rather than its successes bring stigma to the mentally ill. Simply continuing to insist that most of mentally ill are not violent and that the violence caused by the mentally ill is a small fraction of the total violence in the United States will not erase the stigma or calm the anxieties of the public. Hence, it behooves all parties to take active measures to seek to reduce the failures of the system to a minimum. Only in this way is the stigma attached to the mentally ill likely to be eased and the credibility of and public confidence in mental health professionals likely to be enhanced.

Another objection waits. It is that inevitably some number of the competent but violence-prone mentally ill will harm others and be brought to trial for their acts. If they are not able to employ the criminal insanity defense, they may be found guilty of their acts and sent to prison. The very important difficulty is that they will be placed in institutions not equipped to offer them appropriate treatment. It is inappropriate to treat them as, and house them with, ordinary prisoners. It is bad for the mentally ill because they will not receive the treatment they need and bad for ordinary prisoners because they will be housed with those who are unstable and possibly violent. However, the matter is complicated by the fact that, as a number of studies have demonstrated, a considerable number of ordinary prisoners show distinct symptoms of mental illness or have been treated for mental illness in the past. Hence, there is a considerable number of prisoners who are in need of treatment.[39] The proper solution is simple and obvious: Adequate mental health treatment facilities should be established in prisons, just as prisons have facilities to treat physical maladies. In some cases, it is likely that separate units designed to address the needs of mentally ill prisoners will be most suitable. If so, the proper response is to construct them.

The final objection is perhaps the most troubling and the least tractable. It is that the United States has proven unwilling to provide adequate funding and support for mental health treatment in the past and is unlikely to do so at any time in the future.[40] Because this is so, extra funds are unlikely to become available to address the needs of, and the threats posed by, the mentally ill who are potentially dangerous. Programs established to address these problems would therefore only stretch a threadbare system yet further, deplete its inadequate fund of resources, and, worse, remove resources from equally deserving but nonviolent patients. This is an important and troubling difficulty and it should not be overlooked or belittled. However, concerns of public safety and the welfare of potential victims do matter. One thousand homicides, although small in relation to the total, is a considerable number. Further, the number of victims of nonlethal but harmful violence is likely to be several times larger than the number of homicides. Mental health professionals, as do all citizens, share the obligation to enhance public safety. In fact, their obligation is weightier than that of ordinary citizens because they are able to undertake policies to reduce the potential for violent behavior by their patients. In addition, successful programs aimed at reducing the violence of mental health patients would both increase the public's confidence in mental health professionals and the institutions they operate and reduce the stigma of violence that attaches to all the mentally ill and not simply those who are potentially violent.

Conclusion

It is true that the mentally ill whose symptoms are active are more violent than the population as a whole, and their propensity to violence is significantly enhanced if they also engage in substance abuse. It is also the case that mental health professionals have the means to identify which groups of mentally ill people are most apt to become violent. Finally, outpatient commitment programs are not necessarily ineffectual. However, these matters of fact do not suffice to justify overriding the right to consent to treatment enjoyed by mentally ill in order to enhance public safety. Overriding is unjustified because it is inequitable. The examples of automobile travel and possession of firearms demonstrate that the American population is unwilling to accept restrictions on its own activity equivalent to those it would impose on the mentally ill in order to bring about a greater gain for public safety.

The above does not imply that the public has no options available for responding to the threat posed by the potentially violent mentally ill or that the public has no responsibility in this matter. It certainly can and ought to support assertive management programs that will allow the competent mentally ill to reside in the community but under effectual supervision. Further, the public and mental health professionals are obliged to devise machinery to seek to identify those groups of mentally ill who are prone to violence and make opportunities for treatment available to them. Those mentally ill who are competent to give consent to treatment but do not should be placed in the same legal category as the voluntarily intoxicated; that is, they should be held responsible for their condition and legally accountable for acts performed while under its influence.

Notes and References

[1]McFadden, R. D. (1999) New York nightmare kills a dreamer. *New York Times*, January 5, p. A1; Cooper, M. (1999) Suspect has a history of mental illness, but not of violence. *New York Times*, January 5, p. B6.

[2]Winerip, M. (1999) Report faults care of man who pushed woman onto tracks. *New York Times*, November 5, p. B1. *See also* The New York State Commission on Quality of Care for the Mentally Disabled and The Mental Hygiene Medical Review Board (1999) In the matter of David Dix ("David Dix" is a pseudonym for Andrew Goldstein). November. Downloaded July 17, 2000 <http://www.cqc.state.ny.us/dix.htm>

[3]Kleinfield, N. R., with Roane, K. R. (1999) Subway killing casts light on suspect's inner torment," *New York Times*, January 11, p. A1; Winerip, M. (1999) Bedlam on the streets, *New York Times Mag.*, May 23, pp. 42–49, 56, 65–66, 70.

[4]Rohde, D. Subway killer's defense cites past attacks. *New York Times*, October 19, p. B3.

[5]Winerip, M. Bedlam on the streets, p. 45.

[6]Goode, E. (1999) Experts say state mental health system defies easy repair. *New York Times*, November 14, p. B39; Winerip, M., Bedlam on the streets, p. 70; New York State Bill A08477.

Downloaded April 5, 2000 <http://www.stopabuse.net/ny/BILLA08477.htm> and Office of the Governor of New York (1999) Governor's 'Kendra's Law' to protect public mentally ill, Press Release, May 19. Downloaded April 4, 2000 <http://www.state.ny.us/governor/press/year99/may19_99.htm>.

[7]Bazelon Center for Mental Health Law (2000) Involuntary outpatient commitment. May 26. Downloaded June 14, 2000 <http://www.bazelon. org/iocpage.html> and Bazelon Center for Mental Health Law (1999) State involuntary outpatient commitment laws. Downloaded June 14, 2000 <http://www.bazelon.org/ioccchartintro.html>.

[8]Winerip, M. Bedlam on the streets, pp. 44, 70.

[9]Coid, J. W., Taylor, P. J., and Monahan, J. (1996) Dangerous patients with mental illness: increased risks warrant new policies. *Bri. Med. J.* **312,** p. 967.

[10]McCubbin, M. and Cohen, D. (2000) Subject: analysis of the scientific grounds for forced treatment, Open letter to the Little Hoover Commission, State of California, February 1, p. 3. Downloaded June 14, 2000. <http.:www.connix.com/~narpa/cal_ioc.htm>.

[11]Several studies conclude that, under certain conditions, patients assigned to involuntary outpatient commitment programs fare better than those who enter outpatient treatment programs voluntarily. *See* Swanson, J. W., Swartz, M.S., Borum, R., et al. (2000) Involuntary outpatient commitment and reduction of violent behavior in persons with severe mental illness. *Brit. J. Psychiatry* **176,** 324–331; Zanni, G. and deVeau, L. (1986) Inpatient stays before and after outpatient commitment (in Washington, D.C.). *Hospital and Community Psychiatry* **37,** 941–942; Munetz, M. R., Grande, T., Kleist, J. et al. (1996) The effectiveness of outpatient civil commitment. *Psychiatr. Serv.* **47,** 1251–1253; Rohland, B. (1998) The role of outpatient commitment in the management of persons with schizophrenia. Iowa Consortium for Mental Health Services, Training, and Research, May. One widely cited study found no benefit from involuntary outpatient commitment programs: Policy Research Associates, (1998) Final report: research study of the New York City Involuntary Outpatient Commitment Pilot Program (at Bellevue Hospital). Policy Research Associates, Inc., Home Page, December 4. Downloaded June 14, 2000. <http://www.prainc.com/IOPT/opt_toc.htm>.

Although the Policy Research Associates stress the limited scope of their conclusions and the limitations that beset the Bellevue program, their study is cited by several groups who are eager to assert that such programs are no more effective than noncoercive programs. *See* Bazelon Center for Mental Health Law (2000) Studies of outpatient commitment are misused. June 13. Downloaded June 14, 2000. <http://www.bazelon.org/opsctud.html>; and McCubbin, M. and Cohen, D. Subject: analysis of the scientific grounds for forced treatment.

[12]Copeland, R. E. Commissioner, Vermont Department of Developmental and Mental Health Services (1999–2000) Vermont's vision of a public system for developmental and mental health services without coercion. Fall–Winter pp. 6–9. Downloaded June 14, 2000 <http://www.state.vt/dmh>.

[13]Winerip, M. Bedlam on the streets, p. 70. *See also* International Association of Psychosocial Rehabilitation Services (1999) Background position statement on involuntary outpatient commitment. Approved: November by the IAPSRS Board of Directors. Downloaded June 23, 2000. <http://www.iapsrs.org/pubs-ioc_long.htm>.

[14]Marzuk, P. M., (1996) Violence, crime, and mental illness: how strong a link? *Arch. Gen. Psychiatry* **53,** p. 2; Anon. (2001) Violence and mental health—Part I. *Harvard Mental Health Let.* **16,** p. 1.

[15]Soyka, M. (2000) Substance misuse, psychiatric disorder and violent and disturbed behavior. *Brit. J. Psychiatry* **176,** 345–350; and Steadman, H. J., Mulvey, E. P., Monahan, J. et al. (1998) Violence by people discharged from acute psychiatric inpatient facilities and by others in the same neighborhoods. *Arch. Gen. Psychiatry* **55,** 393–401. *See also* American Psychiatric Association (1996) Violence and mental illness. APA Online: Public Information. January 9. Downloaded June 14, 2000 <http://www.psych.org/public_info/VIOLEN~1.HTM>; Coid, J. W., Taylor, P. J., and Monahan, J. Dangerous patients with mental illness: increased risks warrant new policies, p. 965; Anon. Violence and mental health—Part I, p. 2; Mullen, P. E. (2000) Forensic mental health. *Brit. J. Psychiatry* **176,** 307–311; Brennan, P. A., Mednick, S. A., Hodgins, S., et al. (2000) Major mental disorders and criminal violence in a Danish birth cohort. *Arch. Gen. Psychiatry* **57,** 494–500.

[16]Coid, J. W., Taylor, P. J., and Monahan, J. Dangerous patients with mental illness: increased risks warrant new policies, p. 966; Coid, J. W. (1996) Dangerous patients with mental illness: increased risks warrant new policies, adequate resources, and appropriate legislation, *Brit. Med. J.* **312,** 965.

[17]Steadman, H. J., Mulvey, E. P., Monahan, J., et al. Violence by people discharged from acute psychiatric inpatient facilities and by others in the same neighborhoods, pp. 393–401; Steadman, H. J., Mulvey, E. P., Monahan, J., et al. (1998) Response to the *National Review. MadNation* Front Page, July 14. Downloaded June 14, 2000. <http://www.madnation.org/citations/macarthur.htm>; Dyer, C. (1996) Violence may be predicted among psychiatric patients. *Brit. Med. J.* **313,** 318; Mullen, P. E. (1999) Forensic Mental Health; Executive Summary, The MacArthur Violence Risk Assessment Study (1), The MacArthur Research Network on Mental Health and the Law, April. Downloaded June 14, 2000 <http://ness. syst.Virginia.EDU/macarthur/violence.html; Brennan, P. A., Mednick, S. A., Hodgins, S., et al. Major mental disorders and criminal violence in a Danish birth cohort.

[18]The New York State Commission on Quality of Care for the Mentally Disabled and The Mental Hygiene Medical Review Board, In the matter of David Dix, p. 10; Winerip, M., Bedlam on the streets, p. 46. For more general conclusions about the mentally ill, *see* Coid, J. W., Taylor, P. J., and Monahan, J., Dangerous patients with mental illness: increased risks warrant new policies, p. 966; American Psychiatric Association (1996), Violence and mental illness, APA Online: Public Information, January 9. p. 5. Downloaded June 14, 2000 http://www.psych.org/public_info/VIOLEN~1. HTM; Taylor, P. J. and Monahan, J. (1996) Commentary: dangerous patients or dangerous diseases? *Brit. Med. J.* **312,** 967–969.

[19]Marzuk, P. M. Violence, crime, and mental illness: how strong a link? pp. 481–486.

[20]This is the belief of several prominent researchers in the field. *See* Steadman, H. J., Mulvey, E. P., Monahan, J., et al. (1998) Response to the *Nat. Rev.* 2–3.

[21]American Psychiatric Association (1983) Statement on prediction of dangerousness, March 18. *See also* National Mental Health Asso-

ciation (2000) Constitutional rights and mental illness, p. 2. Downloaded June 14, 2000. <http://www.nmha.org/position/ps072899.cfm>.

[22]Dyer, C. Violence may be predicted among psychiatric patients, p. 318; Executive Summary, The MacArthur Violence Risk Assessment Study (1).

[23]Marzuk, P. M. Violence, crime, and mental illness: how strong a link?; Walker, N. (1994) Dangerousness and mental disorder, in Griffiths, A. P. (ed.), *Philosophy, Psychology and Psychiatry*, Cambridge University Press, Cambridge, pp. 184–185.

[24]Mossman, D. (1994) Assessing predictions of violence: being accurate about accuracy. *J. Consult. and Clini. Psychol.* **62,** 783–792; Taylor, P. J. and Monahan, J. Commentary: dangerous patients or dangerous diseases? p. 998.

[25]Monahan, J., Steadman, H. J., Appelbaum, P. S., et al. (2000) Developing a clinically useful actuarial tool for assessing violence risk, *Brit. J. Psychiatry* **176,** 312–319; Dyer, C. Violence may be predicted among psychiatric patients, p. 318; Coid, J. W., Taylor, P. J., and Monahan, J. Dangerous patients with mental illness: increased risks warrant new policies, adequate resources, and appropriate legislation, pp. 965–966.

[26]*Tarasoff v. Regents of the University of California*, 131 California Reporter 14, decided July 1, 1976.

[27]Wexler, D. B. (1996) Therapeutic jurisprudence in clinical practice, *Am. J. Psychiatry* **153,** 453; McIntosh, D. M., and Cartaya, C. Y. (1992) Psychotherapist as clairvoyant: failing to predict and warn. *Defense Council J.* **59,** 569–573; Gagne, P. (1989) More laws for better medicine? *Am. J. Psychiatry* **146,** 819; Monahan, J. (1993) Limiting therapist exposure to "Tarasoff" Liability. *Am. Psychol.* **48,** 242–249; Anon. Violence and mental health—Part II. *Harvard Mental Health Letter* **16,** pp. 1–4. Monahan, J., Appelbaum, P. S., Mulvey, E. P., et al. (1993) Legal report: ethical and legal duties in conducting research on violence: lessons from the MacArthur Risk Assessment Study. *Violence Victims* **8,** 387–396.

[28]US Supreme Court Justice Robert H. Jackson originated this this widely quoted phrase in his dissent to *Terminiello v. Chicago*, 337 U.S 1, 37 (1949).

[29]McCubbin, M. and Cohen, D. Subject: analysis of the scientific grounds for forced treatment, p. 3; Siegel, N. (2000) Statement of Norman Siegel, Executive Director of the New York Civil Liberties Union Concerning 'Kendra's Law,' The New York Civil Liberties Union Home Page. Downloaded April 4, 2000. http://www.nyclu.org/kendrastmnt.html; International Association of Psychosocial Rehabilitation Services. Background position statement on involuntary outpatient commitment.

[30]The source for the figures on homicides is the United States Bureau of Justice. (2000) Homicide trends in the United States. Bureau of Justice Statistics Home Page. Downloaded July 30, 2000. <http://www.ojp.usdoj. gov/bjs/homicide/totals.txt>. In recent years, homicides have steadily declined from a high of 24,700 in 1991 to a low of 16,910 in 1998. E. Fuller Torrey, psychiatrist and president of the Treatment Advocacy Center, made the assertion that the mentally ill commit about 1000 homicides a year in the United States. His figure is based on a 1988 Department of Justice study that concluded that the mentally ill committed 4.3% of the 20,680 US homicides in 1988. Because the number of homicides in the United States increased steadily in the next few years, Dr. Torrey assumed that the percentage of homicides committed by the mentally ill would remain constant at 4.3 and therefore increase to perhaps 1000. Because the number of homicides decreased to 16,910 in 1998, the corresponding rough estimate for homicides committed by the mentally ill would also decrease to 727. Treatment Advocacy Center. (2000) Briefing papers: violence and severe mental illness. Treatment Advocacy Center Home Page. Downloaded July 30, 2000. <http://www.psychlaws.org/BriefingPapers/BPViolence.htm>.

[31]For example, according to the Centers for Communicable Disease, 95,644 deaths resulted from accidents or unintentional injuries in 1997. Centers for Communicable Diseases. (2000) Accidents/unintentional injuries, *FASTATS A to Z.* Downloaded August 3, 2000. <http://www.cdc.gov/nchs/fastats/acc-inj.htm>. The Federal Aviation Agency reports that, in 1997, airplane accidents resulted in 736 deaths. Federal Aviation Agency, *Aviation Safety Statistical Handbook: 1999 Annual Report,* FAA, Washington, DC, pp. 7–1.

[32]Centers for Communicable Diseases (2000) *Nat. Vital Statist. Rep.* **48.** Downloaded July 30, 2000. <http://www.cdc.gov/nchs/releases/00news/00news/finaldeath98.htm>; Centers for Communicable Diseases. (2000) Accidents/unintentional injuries, *FASTATS A to Z.* Downloaded July 30, 2000. <http://www.cdc.gov/nchs/fastats/acc-inj.htm>.

[33]Centers for Communicable Diseases. (2000) Nat. Vital Statist. Rep. **48.** *See also* National Mental Health Association, Constitutional rights and mental illness.

[34]The Federal Aviation Agency statistics reveal that there were an average of 867 deaths resulting from airplane crashes from 1994 to 1999. For the longer period from 1982 to 1993, there were an average of 906 deaths each year in airplane crashes. Federal Aviation Agency, pp. 7–1 and A-4.

[35]Goode, E. Experts say state mental health system defies easy repair, p. 42; Winerip, M. Report faults care of man who pushed woman onto tracks, p. B6; The New York State Commission on Quality of Care for the Mentally Disabled and The Mental Hygiene Medical Review Board, In the matter of David Dix, pp. 13–14 and International Association of Psychosocial Rehabilitation Services, Background position statement on involuntary outpatient commitment.

[36]Sweitzer, B. G. (1997) Implicit redefinitions, evidentiary proscriptions, and guilty minds: intoxicated wrongdoers after Montana v. Engelhoff. *Univ. Penn. Law Rev.* **146,** 269–322.

[37]International Association of Psychosocial Rehabilitation Services, Background position statement on involuntary outpatient commitment.

[38]Walker, N. Dangerousness and mental disorder, pp. 180–181.

[39]Winerip, M. Bedlam on the streets, p. 46; Marzuk, P. M. Violence, crime, and mental illness: how strong a link?; Anon. Violence and mental health—Part I, p. 1; Gunn, J. (2000) Future directions for treatment in forensic psychiatry. *Brit. J. Psychiatry* **176,** 332–338.

[40]Winerip, M. Bedlam on the streets, p. 70.

Abstract

Can ethical theory inform the choice of intervention in circumstances under which there is more than one effective treatment available? We argue that the philosophical and ethical perspective of existentialism (which emphasizes the autonomy, responsibility, and free choice of the individual) can help answer this question. In our analysis, we selected depression- and anxiety-related problems as examples because they are associated with two alternative intervention approaches of demonstrated efficacy. These alternatives are cognitive behavior therapy and pharmacotherapy. Although both treatments would be considered to be reasonably ethical from a teleological ethical perspective, existential philosophy would suggest that the former has an additional ethical advantage.

Cognitive Behavioral and Pharmacological Interventions for Mood- and Anxiety-Related Problems

An Examination from an Existential Ethical Perspective

David Cruise Malloy and Thomas Hadjistavropoulos

Introduction

In existential ethical philosophy, the autonomy of the individual is considered to be of paramount importance (e.g., ref. 1). We argue that this philosophical and ethical perspective can inform the choice of interventions in a mental health context and

From: *Biomedical Ethics Reviews: Mental Illness and Public Health Care*
Edited by: J. Humber & R. Almeder © Humana Press Inc., Totowa, NJ

have selected for our analysis two types of problems that frequently bring people into contact with providers of mental health services. These problems are depression and anxiety/fear. In this chapter, we focus on the types of intervention that have been shown to be the most effective in alleviating these problems: pharmacotherapy and Cognitive Behavior Therapy (CBT). Both of these approaches can be used to overcome depression (e.g., refs. 2 and 3), obsessions and compulsions, as well as problems with panic (e.g., refs. 4–7). Under such circumstances, where two types of treatment are effective (i.e., CBT vs pharmacotherapy), existentialism can inform us as to which of these two treatments is preferable from an ethical perspective.

Pharmacotherapy and the Medical Model

Pharmacotherapy for mood- and anxiety-related problems can be examined within the context of a broader medical approach that has been traditionally viewing such problems as "illnesses" (e.g., refs. 8–10). The term "disease" or "illness" implies a process that has a specific etiology and predictable outcomes and presents with signs and symptoms.[11] Construing mental disorders as "diseases" has been described as the *medical disease model* or *medical model* of psychopathology (e.g., ref. 12). The basis of this model is derived from a bioscientific approach to objectifiable conditions, namely physical illnesses. Issues relating to the causes and definitions of physical conditions or diseases can be often answered deterministically and objectively.

Child[12] points out that medicine concerns itself with diseases that are, regardless of their etiology, considered to be occurring within the individual patient or client and many healthcare providers often generalize this approach to mental disorders. Similarly, the use of terms such as "patient" and "mental hospital" encourages such disease conceptualizations of mental disorders. Nonetheless, behavioral disorders are largely the result of interactions within social systems that include, but are not limited to, important family and peer relationships.[13] Research shows, for example, that negative life events maintain and precipitate depression (e.g., ref. 14) and that having good social support systems can prevent mental disorders (e.g., ref. 15). Social factors have to act on any genetic predispositions that may exist[16] before patterns of abnormal behavior can emerge. Undermining the importance of any such factors can have ethical implications. Nonetheless, in recent decades, there has been a major shift in psychiatric training with an increased emphasis on biological treatment (e.g., ref. 17) and a relative underemphasis of social factors, habits, and learning. Despite the fact that the vast majority of mental disorders do not meet the criteria for "disease" (e.g., they lack specific and clear-cut etiology), disease conceptions abound. Milton and Wahler,[13] for example, quoted from a 1962 article in *Mind Over Matter*: "If everyone would just realize that mental illness is no different from any other prolonged disease and that a heart attack victim differs only from a mental victim in the localization of the affliction, the psychiatrist's therapist's job would be greatly simplified" (p. 8).

When the medical model was originally being conceptualized, it applied primarily to psychotic behavior and was useful because many institutionalized cases were, in fact, organic in nature and were the result of conditions such as syphilis.[13] Many of these organic conditions, however, have almost disappeared because of developments in medicine. Most contemporary proponents of the medical model do not believe that all mental disorders have organic causes but, nonetheless, use specific diagnoses to explain causes and symptoms as well as to prescribe a treatment. Although such explanations can account for symptoms of physical illnesses (e.g., "X has a runny nose because he has influenza [i.e., was exposed to an influenza virus]"), they are rather *circular* when applied to mental disorders (e.g., "X is not interested in playing golf because he has depression"). There is a conceptual problem involved in the explanation "X behaves in this fashion because he is neurotic." Neurosis is a descriptor of behavior not an explanation for it. Similarly, using disease-like labels to describe people with mental disorders and to construe diagnoses as reflecting discrete categories is usually incorrect. In fact, dimensional views of psychological problems appear to be gaining favor (e.g., ref. 18). Generalized Anxiety Disorder, for example, appears to be an extreme form of the type of trait anxiety that the vast majority of people experience in varying degrees. There is little evidence for the idea that Generalized Anxiety Disorder represents a discrete diagnostic entity. The major features of this phenomenon are observed across all types of anxiety disorder and none of its features are unique or specific to Generalized Anxiety Disorder (e.g., ref. 19).

The term "mental illness" is used widely to describe conditions that have largely (but not exclusively) psychosocial causes such as depression, alcoholism, compulsive gambling, and conduct disorder of childhood. Blume,[20] for example, suggested that the medical model involves an "illness" concept and that compulsive gambling fits this model. Nonetheless, compulsive gambling bears more similarities to socially constructed problems than to medical ones.[21] The medical model has also been criticized for being reductionistic and mechanistic in its essence because it reduces the emotional, interpersonal, and ethical conflicts or conflicts in life in favor of biochemical levels of explanation (e.g., refs. 22 and 23).

In contrast to the medical model, psychosocial and integrated conceptions of mental disorders (that do not deny the notion of genetic predispositions) are based in more egalitarian role relationships between client and practitioner, encourage initiative and autonomy in the client, as well as culturally and individually based treatment.[23,24] Furthermore, they encourage taking action to improve adverse life circumstances (e.g., to find a more satisfying job) as well as to correct self-defeating aspects of behavior (e.g., to help an unassertive person develop assertiveness skills). Treatments designed to treat mental disorders as "diseases" are less likely to encourage people to take active steps to improve their lives' circumstances and to rely on their own psychological resources.

The medical model of disease has several *social consequences*. Specifically, *stigma* or *blame* may be reduced because the victim of a "disease" or patient is not held responsible for his or her ailment because illness is something that happens to the person.[25,26] For example, according to the National Alliance for the Mentally Ill, a US organization with 220,000 members, "Mental illnesses are not the result of personal weakness, lack of character, or poor upbringing. Most importantly, these brain disorders are treatable. As a diabetic takes insulin, most people with serious mental illness need medication to help control symptoms."*

Additionally, as Milton and Wahler[13] point out, in order to become well, the victim is expected to be more or less a *passive recipient of treatment* and to have things done to him or her (e.g., administration of medications). Moreover, the sick or ill person receives special consideration (e.g., an employee may be excused from work with pay). Greenberg and Bailey[27] noted that it has even been argued that the identification of behavior patterns as diseases could be used to support conclusions in legal proceedings that a defendant's illegal act was caused by a condition that could relieve him or her from legal responsibility. Nonetheless, it has also been argued that whether or not mental disorders are biological entities is not really relevant to the notion of responsibility because many human acts (e.g., crime) may have biological causation, yet our society feels that people who commit these acts should still be considered morally responsible.[27] Greenberg and Bailey also suggested that all behavior is ultimately biologically determined and that this does not reduce the degree of responsibility that we assign to people for their actions. This is probably because society recognizes the social factors that play a

*Available at http://www.nami.org/disorder/whatis.html.

role in the determination of most behaviors and accepts the notion that people make choices about their actions.

The medical model implies that pharmacotherapy would be an appropriate treatment of choice for many of the conditions described in the *Diagnostic and Statistical Manual of Mental Disorders* (DSM-IV).[28] Although we agree that pharmacotherapy should be the treatment of choice under many circumstances (e.g., in the treatment of many people who have serious problems relating to bipolar mood), our interest lies in situations in which effective pharmaceutical and nonpharmaceutical treatments are available (e.g., most depression- and anxiety-related problems). We examine whether theoretical ethics can inform us on the treatment of choice under such circumstances.

Pharmacotherapy for depression- and anxiety-related problems most often implies seeing a physician every two weeks (or less often) for a very brief period in order to evaluate whether the prescribed medication(s) is (are) effective. If symptoms are not improved (or if serious side effects are present), one or more of the following are likely courses of action: (1) the medication dosage is adjusted; (2) the "patient" is switched to an alternative medication; and (3) one or more additional medications are added to the prescribed regiment. Blood tests are often used to assess whether the medication levels fall within the "therapeutic range." What appears to be missing in this approach is a sufficient involvement of the "patient" in his or her own therapy other than, of course, the responsibility of drug consumption. Although the chemical adjustments may lead to states that are more within normal limits, the "patient" has not been provided with any nondrug-related strategies that would allow him to improve his or her life's circumstances. The "patient's" state of happiness or anxiety is bound not to personal responsibility, but rather to "taking one's medication."

The extent to which sufficient information about treatment options and alternatives is provided by pharmacotherapists varies substantially across providers. Apart from pharmacotherapy, most of the interventions that have been shown to be effective for anxiety- and depression-related problems fall under the category of cognitive behavior therapy.[29]

Cognitive Behavior Therapy

Cognitive Behavior Therapy (CBT) can be construed as falling within a larger "learning therapies tradition" that includes behavioral, cognitive, and social learning perspectives. The main assumption is that problems related to anxiety and depression are largely the result of past learning and relate, at least in part, to beliefs and attitudes that people have about themselves, the future, and the world around them.[30,31] People with depression often express beliefs such as "I am worthless" and "I am a total failure." They are also characterized by a strong sense of hopelessness about the future.[32] Anxious people, on the other hand, are believed to overestimate risk or the probability of danger and misperceive safe situations as being dangerous.[31] They tend to avoid such situations and/or endure them with fear and apprehension.

Although the role of past learning, beliefs, and attitudes are at the fountain top of the cognitive behavioral explanation of anxiety and depression problems, cognitive behavior therapists do not deny the role of biological and genetic propensities (e.g., ref. 30). Nonetheless, they believe that the individual can overcome depression and anxiety by objectively examining his or her belief system and by working toward changing any views that are not supported by evidence. Fear-related problems can be overcome when the person finds the courage to enter situations that were previously avoided (because of fear) and to stay in such situations until the fear extinguishes.[33] Such approaches have shown great success in alleviating suffering associated with depression and anxiety.

In order to examine some of the ethical implications of the cognitive behavioral approach, we will outline some fundamental characteristics of the therapy. In the initial stages, the therapist is not primarily interested in diagnosing the client but in identifying the issues that are of greatest concern to him or her (e.g., sadness, hopelessness, loss of interest, fear of travel, fear of death). The goals and expectations of the client regarding the therapy are also outlined and the goals of the intervention are agreed upon. During the initial stages of treatment, the therapist will typically clarify that the client is unlikely to demonstrate the changes that he or she desires merely by attending the weekly CBT sessions. Instead, the client must work and practice assignments (designed in collaboration with the therapist) outside the session. Such assignments involve, for example, practicing certain skills (e.g., assertiveness) and keeping records of his or her upsetting thoughts. This type of work is considered essential and of paramount importance to overcoming the client's problems. Each session has a clearly defined, mutually agreed upon agenda of topics to be covered and both the therapist and client contribute items. The therapist engages the client in "collaborative empiricism"[30] during which the client's beliefs are discussed in a Socratic fashion. The therapist will typically not provide the client with answers but will encourage him or her to reach her or his own solutions through Socratic dialogue (e.g., "You told me that when you are upset, you either blow up or you say absolutely nothing. But is there another way? Is there another way that could help change the situation that is upsetting you?").

The types of assignment designed during therapy (often referred to as "homework assignments") can have the form of informal experiments designed to help the client find the truth. A person who believes that no coworker would ever agree to have coffee with him or her may ask several coworkers to go for coffee in an attempt to determine whether his belief is correct. This strategy is in stark contrast to pharmacotherapy in terms of its explicit attempt to provide the individual with specific skills that empower autonomous decision making. The individual is not a passive recipient of a chemically-based therapy, but rather an active and responsible participant in a reason-based intervention.

Ethical Theory and Treatment Selection

One school of thought that can be used to answer ethical questions is consequentialism, more formally known as teleology.[34,35] Within the teleological tradition, the consequences of a behavior are used to determine whether or not this behavior is ethical. Therapy X is chosen over therapy Y if the former is more effective than the latter. When two mental health interventions are both approximately equally effective in reducing human suffering, teleology is not useful in making an ethical determination as to whether one of the treatments is preferable over the other. Issues pertaining to monetary cost of treatment may have some teleological relevance, but they are rather peripheral when compared to the ultimate concern, which is the reduction of human suffering.

The other traditional school of thought used to resolve ethical issues is termed nonconsequentialism or deontology.[36] Deontology is not focused upon the ends of action but upon the means. In other words, the individual has a duty to follow principles, laws, and codes. For example, health care professionals are duty-bound to follow the principles and codes adopted by their professional associations. Although professional codes of ethics articulate ethical principles, they do not provide the direction needed when choosing between two effective forms of therapy, which both accord with these principles. Because neither teleology nor deontology can answer the question regarding which of two effective, rule following treatments is ethically preferable, we turn to another perspective of ethical philosophy: existentialism.

Existentialism

Existentialism has evolved as a philosophical criticism of and revolt against both teleological and deontological camps.

Despite its many variations, existentialism views the capacity for free and autonomous choice as fundamental to the essence of the human being. This being the case, any encroachment upon one's ability to be autonomous circumvents the goal of authenticity and has negative ethical implications. From an existential standpoint, approaches that emphasize authenticity, responsibility for the individual, and free choice are more ethical than those that do not.[1,37,38] In the following subsections, we elaborate on the most fundamental components of existentialism and examine them in relation to pharmacotherapy and CBT.

Ontology

As one might expect of a philosophy of individualism, a considerable array of perspectives among existential thinkers exists. Despite its rather eclectic nature, there is an accepted ontological view of the individual that unifies this philosophy and its proponents. Sartre[1] has perhaps most succinctly described the existential ontology or the nature of the human being in his famous phrase "existence precedes essence" (p. 13). This statement implies that the human exists first and foremost in the world and then each decision he or she makes forms his or her essence; or as Sartre himself suggests,

> Man exists, turns up, appears on the scene, and only afterwards, defines himself. If man, as the existentialist conceives him, is indefinable, it is because at first he is nothing. Only afterward will he become something, and he himself will have made what he will be. . . . Not only is man what he conceives himself to be, but he is also only what he wills himself to be after this thrust toward existence. (p. 15)

The personal development or self-determination of essence is possible because the individual has the freedom to make decisions (i.e., essence is not predetermined by divine or societal intervention). Although it is true that the individual is bound by certain facts or by what Heidegger[39] describes as "facticity" (e.g., our genetic makeup), most decisions are within our capacity to make. Macquarrie[37] suggests that "it is the exercise of freedom and the ability to shape the future that distinguishes man from all other beings that we know on earth. It is through free and responsible decisions that man becomes authentically himself" (p. 16). Sartre[1] exposes this notion of determinism and self-determinism in the following dialectic:

> If you're born cowardly, you may set your mind perfectly at rest; there's nothing you can do about it; you'll be cowardly all your life, whatever you may do. If you're born a hero, you may set your mind just as much at rest; you'll be a hero all your life; you'll drink like a hero and eat like a hero. What the existentialist says is that the coward makes himself cowardly, that the hero makes himself heroic. There's always a possibility for the coward not to be cowardly any more and for the hero to stop being heroic. What counts is total involvement. . . . (p. 35)

This kind of involvement, that the therapist seeks in CBT, appears to be considerably lacking in pharmacotherapeutic strategies.

As we have discussed, the existential view of the human being's essence is that we exist first and then we become who we are. Who we are is a function of the decisions we make and the responsibility we take for all of our actions (i.e., "good faith"). To make decisions in "bad faith" is to allow our control over our essence to be taken on by others. In other words, if we adopt or conform to a position or a role (e.g., the passive patient) that is dictated by the will of others as opposed to our own authentic will, we act in "bad faith"; that is, we relinquish our potential to develop into unique, actualized, and authentic individuals. In addition, we relinquish our responsibility for behavior to the health care provider, to medication, or to significant others.

The medical model of mental disorders implies that the "patient" is not only a bystander in terms of the onset of depression but also with regard to the treatment of the "diseased" state.

For example, Milton and Wahler[13] described the medical model in the following: "Human disorders are conceived, like other phenomena to be outgrowths of naturalistic processes occurring in orderly if complex sequences" (p. 247). Further, they argue that there is, in some respects, a common theme between contemporary scientific and medieval supernatural explanations of mental disorders. Both of these perspectives present the individual as the passive host of a biological or spiritual end state. The individual is not responsible for the onset or the termination of such a state and must endure external treatment or exorcism in order to become well again.

Cognitive behavior therapy provides a different ontological view of the individual. Treatment is not geared toward the biological "patient" but toward the rational individual who can be self-directed and an active participant in his or her own healing. This view coalesces with the existential perspective of the individual as one who is in the process of becoming who he or she will be. In addition to a generally agreed on ontology based on the freedom to choose personal destiny, existential philosophers also share at least two additional themes in their writings. These are responsibility and an intense suspicion of societal attempts to normalize behavior.

Responsibility and Existential Seriousness

According to existentialism, the individual is free to choose a personal destiny (e.g., to become a hero or coward). With this choice, however, responsibility must be accepted. It is the weight of this responsibility that is the basis for the so-called existential anxiety, angst, anguish, and torment. It is important to note that although existentialists choose a personal destiny, they are responsible for the impact that their decisions have not only on themselves but also on all of humanity. Sartre[1] states that

> . . . existentialism's first move is to make every man aware of what he is and to make the full responsibility of his exis-

> tence rest on him. And when we say that a man is responsible for himself, we do not only mean that he is responsible for his own individuality, but that he is responsible for all men. . . . Thus our responsibility is much greater than we might have supposed, because it involves all mankind. (pp. 16–17)

Failing to take responsibility for one's own actions and the impact that these actions may have on others is what Sartre[1] terms "acting in bad faith." Kierkegaard[40] speaks at length about the tendency of individuals to hide behind the group (or the family or society) when making a "decision" in order to avoid accountability (i.e., "acting in bad faith"). For example, he states that "a crowd in its very concept is the untruth, by reason of the fact that it renders the individual completely impenitent and irresponsible, or at least weakens his sense of responsibility by reducing it to a fraction" (p. 95).

Proponents of the medical model will often attribute a depressed person's problems to a "chemical imbalance," thus completely removing the responsibility from the person. May and Yalom[41] also comment on the tendency in the realm of the Freudian approach to view the "patient" as determined by instincts and subconscious drives. They state the following:

> The cellular view of the unconscious leads patients in therapy to avoid such responsibility for their actions by such phrases as, "My unconscious did it not I." Existentialists always insist that the patient in therapy accepts responsibility for what he or she does . . . (p. 361)

The existential individual is profoundly responsible for his or her behavior. Blame or praise cannot be directed toward significant others or society-at-large. Authentic individuals are accountable for the circumstances in which they find themselves and for their future essence. Therefore, clients would be viewed as responsible for their current state of mental health and for any intervention that may be recommended through counseling.

We do not believe that the existential model is relevant to all of the problems listed in the DSM-IV. Evidence has linked a variety of conditions such as attention deficit and hyperactivity problems to factors (e.g., central nervous system abnormalities[42]) for which we cannot hold the individual responsible. Nonetheless, in many cases of psychosocial problems, people do make choices. A person who is fearful of air travel may chose to avoid flying or may chose to fly as much as possible as a means of overcoming his or her fear. A depressed person ultimately has the choice of taking certain actions that could improve his or her life's circumstances. In contrast to such psychosocial views, the medical model has the tendency to direct responsibility for any pathology to biological features. The individual becomes the passive host of a diseased state as opposed to an active participant in the aetiology and "cure." Psychosocial approaches aim to empower the client. A person who is extremely fearful and avoidant of specific situations (e.g., heights) is encouraged to find the strength to gradually expose himself or herself to these until he or she gets used to them,[33] or until graduated exposure becomes boring because the fear has been overcome. Such mastery-based exposure approaches emphasize the courage of the individual[33] and his or her ability to gain a sense of self-efficacy.[43]

Society's Leveling of the Individual

The "leveling" or conformity of the individual to the norms of the group or of society is of great concern to existentialist writers. Kierkegaard,[44] the father of existentialism, wrote at length of the leveling effect of the "public." In *The Present Age,* he argued that leveling is a process that destroys the notion of individual *qua* individual. It intends to make the individual a part of the whole. Kierkegaard[44] describes this process in the following:

> A demon is called up over whom no individual has any power, and though the very abstraction of levelling gives the individual a momentary, selfish kind of enjoyment, he is at the same time signing the warrant for his own doom. Enthusiasm may end in disaster, but levelling is eo ipso the destruction of the individual. . . . The abstract levelling process, that self-combustion of the human race, produced by the friction which arises when the individual ceases to exist as singled out by religion, is bound to continue, like a trade wind, and consume everything. (pp. 54–55)

Although this notion of leveling is often directed to the pressure one experiences to conform to societal norms, it can also relate to the tendency for the individual to be perceived empirically and historically as opposed to existentially (i.e., Heidegger's facticity vs dasein). In other words, the individual is observed, studied, or diagnosed as a collection of empirical facts common to the species.[38] This view, Macquarrie[37] argues, is in contrast to the existential perspective that "marks a new transition from the interpretation of knowledge as objectification to understanding it as participation, union with the subjective matter, and entering into cooperation with it" (p. 136).

Buber[45] speaks to this issue in his seminal work, *I and Thou*. He sees two possible relationships that one may have with another. The first is an "I-It" relationship in which the "I" perceives the other as an empirical "thing" or an "It." In this circumstance, the other is not viewed as an existential being in the process or flux of being or experiencing, but rather as a static factual entity. Thus, when another human is perceived as an "It," he or she is leveled into the common empirical and objective realm (i.e., patient as a host of pathology); perceived as "Thou," he or she is understood subjectively and in terms of mutual participation in being (i.e., person as client).

Heidegger[39] expresses concern for the inclination of society and science to objectify the individual. He argues that traditional philosophy (and subsequently science and medicine) has focused upon the calculative study of the essence of beings. Calculative thinking is concerned with logic, utility, description, categorization, focus, and purpose. Although this mode of cognition is important it falls short of a comprehensive understanding of essence because it does not address beings comprehensively.

According to Heidegger,[46] a different kind of thinking (i.e., other than calculative) is required in our efforts to understand human beings. He terms this mode of thinking "meditative" or "poetic." Meditative thinking is characterized by the terms "releasement" and "openness." Releasement refers to a need to focus not only on the main object that is being observed (e.g., the patient) but also on the background that surrounds and helps define it. Openness refers to having an unbiased perspective and to being open to new ideas and interpretations.

Although calculative thinking is necessary for any of us to operate in the mainstream "everydayness" of Western society, we will not come to understand ourselves or others existentially if we do not think meditatively. Heidegger[46] states the following:

> Meditative thinking demands of us not to cling one-sidedly to a single idea, nor to run down a one-way track course of ideas. Meditative thinking demands of us that we engage ourselves with what at first sight does not go together at all. (p. 53)

Meditative thinking leads to an understanding of oneself as *dasein. Dasein* is translated from German to "being there." This implies that the individual is not the static or stagnant sum of his or her attributes, characteristics, or pathologies, but rather he or she is in flux. Another way of describing dasein is to suggest that humans are their mode of experience rather than what their traits or biological characteristics indicate.

The process of leveling in the medical model has the tendency to categorize and to generalize (i.e., Heidegger's conception of "calculative thinking"). This results in the loss of individual identity in favor of group identity. The client is, therefore, not described in terms of his or her particular circumstances, but is diagnosed in terms of the common features of the "disease" or pathology (i.e., Buber's conception of "It"). In contrast, psychosocial conceptions of mental disorders do not only examine problem behaviors but also the person's environment as well as his or her unique strengths and goals. Answers relating to the client's concerns are obtained by collaboratively designing and carrying out informal experiments that can lead the client to a truth that specifically applies to his or her life more than to the life of any other individual. If I want to know whether my belief that no coworker would ever have coffee with me is true, I may have to find out by asking several of them if they would go to the cafeteria with me to have a cup of coffee. Nonetheless, some categorization of individuals can occur within the context of CBT (e.g., diagnosis, type of cognitive errors held by the person [i.e., beliefs that are contradicted by the empirical evidence]). In contrast to medical treatments for depression- and anxiety-related problems, however, diagnostic labels or the categorization of a belief are not necessary preconditions for treatment to proceed and many cognitive behavior therapists intentionally avoid such categorizations. It must also be noted that treatment manuals for CBT exist (e.g., ref. 30). Although any form of manualization can be viewed as reducing free choice and individuality, treatment manuals are essentially outlines within which the client's specific goals and concerns are discussed and addressed in individualized ways. Experienced therapists often tend to deviate from treatment manuals and adopt highly individualized approaches (*see,* for instance, a related discussion by McMullen[47]).

Authenticity

The goal of the existentialist is to be authentic. This implies that one must act in "good faith"; that is, one must choose to acknowledge free will and to take responsibility for all behavior including those decisions that may influence one's mood and level of anxiety. This perspective is manifested when the client is the final arbiter of all decisions involving his or her well-being. This climate of choice is not provided in an environment in which the mental health professional assumes implicitly the posture of knowing and choosing what is best for the patient or when biochemical reductionist strategies are employed rather than those of reasoned action. As stated earlier, Beck, in his approach to cognitive behavior therapy, advocated "collaborative empiricism," a process through which the client and therapist work together in an effort to uncover beliefs and values that may be behind a person's problem and to identify solutions that could lead to overcoming the problem.[30] It is well known that when the client develops the solutions, these are far more likely to be implemented than when the solutions are developed by the therapist.

Conclusion

In this chapter, we have endeavored to explore mental health care through the lenses of existential philosophy. We chose this perspective because the traditional teleological and deontological perspectives seemed to be deficient in defining a strategy that focuses upon the individual as an autonomous being. Furthermore, teleology and deontology do not seem to inform ethical treatment selection in situations in which two or more different treatments are approximately equally effective and in accord with professional codes of conduct.

We discussed two alternative interventions that are pre-eminent in the treatment of mood- and anxiety-related problems. Our examination was based on the existential themes of ontology, responsibility, leveling, and authenticity. Table 1 provides a summary of these themes, as existential tenets are juxtaposed with the two therapeutic options. We conclude that strategies that incorporate the full participation and active involvement of the client contribute to existential autonomy. In contrast, those interventions that tend to reduce depressed and anxious "patients" to passive recipients of therapy or to passive victims of their biology do not measure up to the existential standards of freedom, responsibility, and authenticity.

Although we conclude that CBT approaches are preferable over pharmacotherapy (in situations in which both treatments are considered to be effective) because they are more likely to embrace existential ethical values (e.g., responsibility, authenticity, and free choice), we acknowledge that these approaches are not always perfectly consistent with existential values. This is, in part, because some categorization of individuals can occur within the context of CBT. Although psychotherapies that adopt more existential values than CBT are available (e.g., ref. 41), these have yet to demonstrate their usefulness clearly (e.g., ref. 29). Despite some discrepancies from existential values, CBT embraces such values to a greater extent than does pharmacotherapy, and when either treatment is clinically appropriate (as is the case with the majority of anxiety- and depression-related problems), CBT has an ethical advantage.*

Acknowledgment

The preparation of the chapter was supported in part by a Social Sciences and Humanities Research Council of Canada grant to D.C. Malloy and T. Hadjistavropoulos.

*As teleological ethical philosophy would indicate, competently conducted pharmacotherapy and CBT are both reasonably ethical options in helping persons with many types of depression- and anxiety-related problems.

Table 1
Basic Tenets of Existential, Pharmaceutical, and Cognitive Behavioral Perspectives as They Relate to the Treatment of Depression and Anxiety Problems

	Existentialism	Cognitive Behavioral Perspective	Pharmacotherapy
Ontology	The person is viewed subjectively as a unique individual in the process of becoming.	The client is viewed as an individual who has learned certain beliefs and behaviors. Choices are guided by such beliefs.	The patient is viewed as the host of a pathology.
Responsibility	Individuals take responsibility for all decision-making behavior for themselves and for all of humanity.[1]	The client participates and takes responsibility in the therapy both through "collaborative empiricism" and through participation in the development of courses of action that can help with the achievement of his or her goals.	The medical system assumes responsibility for the patient who is perceived as being unable or lacking the knowledge to take appropriate action.

Leveling	The individual avoids acting in terms of predetermined and inauthentic roles established by society (i.e., acting in "bad faith")	Although categorical descriptors are sometimes used (e.g., type of cognitive error), treatment focuses on specific problems and concerns that are unique to the individual.	The patient is "leveled" into a category of disease.
Authenticity	The individual strives to act in "good faith." Choices are made not from the pressure to conform ("bad faith") but from the position of individually based free will and responsibility.	The client participates in the design strategies that could prove helpful and examines the validity of his or her beliefs. He or she finds the courage to face fears that stand in the way of his or her ability to enrich his or her life's experience.	The patient's authenticity is not of concern to the medical perspective because it is the pathology that is being treated.

References

[1]Sartre, J. P. (1957) *Existentialism and human emotions,* The Winston Library, New York.

[2]Jacobson, N. S. and Hollon, S. D. (1996) Cognitive-behavior therapy versus pharmacotherapy: Now that the jury returned its verdict, it's time to present the rest of the evidence. *J. Consult. Clin. Psychol.* **64,** 74–80.

[3]Rush, A. J., Beck, A. T., Kovacs, M., and Hollon, S. D. (1982) Comparative efficacy of cognitive therapy and pharmacotherapy in the treatment of depressed outpatients. *Cogn. Ther. and Res.* **1,** 17–39.

[4]Ballenger, J. C., Burrows, G. D., Dupond, R. L., Lesser, I. M., Noyes, R., and Pecknold, J. C. (1988) Alprazolam in panic disorder and agoraphobia: results from a multi center trial: I. Efficacy in short-term treatment. *Arch. Gen. Psychiatry* **45,** 413–422.

[5]Barlow, D. H. and Lehman, C. L. (1996) Advances in the psychosocial treatment of anxiety disorders: implications for national health care. *Arch. Gen. Psychiatry* **53,** 727–735.

[6]Craske, M. G., Brown, T. A., and Barlow, D. H. (1991) Behavioral treatment of panic disorder: A two year follow up. *Behav. Ther.* **22,** 289–304.

[7]Zohar, J., Judge, R. and the OCD paroxetine study investigators. (1996). Paroxetine vs. clomipremine in the treatment of obsessive compulsive disorder. *Brit. J. Psychiatry* **169,** 468–474.

[8]Flentge, F., van den Berg, M. D., Bouhuys, A. L., and The, H. T. (2000) Increase in NK-T cells in aged depressed patients not treated with antidepressant drugs. *Biol. Psychiatry* **48,** 1024–1027.

[9]Okasha, A., Lotaief, F., Ashour, A. M., el Mahalawy, N., Seif el Dwala, A., and el-Kholy, G. (2000) The prevalence of obsessive compulsive symptoms in a sample of Egyptian psychiatric patients. Encephale **26,** 1–10.

[10]Ononye, F., Sijuola, O. A., Chukwuani, C. M., Mume, O. C. and Makanjuola, R. O. (2000) Open comparative randomised study of Moclobemide versus Amitryptiline in major depressive illness (DSM-IIIR) in Nigeria. *West Afr. J. Med.* **19,** 148–153.

[11]Hornstra, R. (1962) The psychiatric hospital and the community. Paper presented at the Annual Workshop in Community Mental Health, Pisgah View Ranch, Candler, NC.

[12]Child, N. (2000) The limits of the medical model in child psychiatry. *Clin. Child Psychol. and Psychiatry* **5,** 1359–1045.

[13]Milton, O. and Wahler, R. G. (1969). Perspectives and trends, in *Behavior Disorders: Perspectives and Trends,* 2nd ed., Milton, O. and Wahler, R. G., eds., JB. Lippincott, Philadelphia, pp. 3–16.

[14]Kessler, R. C. (1997) The effects of stressful life events on depression. *Annu. Rev. Psychol.* **48,** 191–214.

[15]Brown, G. W. and Harris, T. O. (1978) *Social Origins of Depression: A Study of Psychiatric Disorder in Women,* Tavistock, London.

[16]Durand, V. M. and Barlow, D. H. (2000) *Abnormal Psychology,* Wadsworth/Thompson Learning, Belmont, CA.

[17]Guze, B. M. (1992) *Why Psychiatry Is a Branch of Medicine,* Oxford University Press, New York.

[18]Widiger, T. and Costa, P. T. (1994) Personality and personality disorders. *J. Abnormal Psychol.* **103,** 78–91.

[19]Barlow, D. H. (1988) *Anxiety and Its Disorders: The Nature and Treatment of Anxiety and Panic,* Guilford, New York.

[20]Blume, S. (1987) Compulsive gambling and the medical model. *J. Gambling Behav.* **3,** 237–247.

[21]Wedgeworth, R. L. (1998) The reification of the "pathological gambler": an analysis of gambling treatment and the application of the medical model to problem gambling. *Perspect. Psychiatr. Care* **34(2),** 5–9.

[22]Sarbin, T. R. and Mancuso, J. C. (1980) *Schizophrenia: Medical Diagnosis or Moral Verdict,* Pergamon, Elmsford, NY.

[23]Wyatt, R. C. and Livson, N. (1994) The not so great divide? Psychologists and psychiatrists take stands on the medical and psychosocial model of mental illness. *Profess. Psychol.: Res. Pract.* **25,** 120–131.

[24]Sharma, S. L., ed. (1975) *The Medical Model of Mental Illness,* Majestic, Woodland CA.

[25]Burnsten, B. (1981) Rallying around the medical model. *Hosp. Community Psychiatry* **32,** 371.

[26]Siegler, M. and Osmond, H. (1974) *Models of Madness, Models of Medicine,* Harper Collins, New York.

[27]Greenberg, A. S. and Bailey, J. M. (1994) The irrelevance of the medical model of law and illness to law and ethics. *Int. J. Law Psychiatry* **17,** 153–173.

[28]American Psychiatric Association (1994) Diagnostic and Statistical Manual of Mental Disorders. 4th ed., American Psychiatric Association. Washington, DC

[29]Hunsley, J., Dobson, K. S., Johnston, C., and Mikail, S. F. (1999) Empirically supported treatments in psychology: implications for Canadian Professional Psychology. *Can. Psychol.* **40,** 289–302.

[30]Beck, A. T., Rush, A. J., Shaw, B. F., and Emery, G. (1979) *Cognitive Therapy of Depression,* Guilford, New York.

[31]Rachman, S. (1998) *Anxiety* Psychology Press, East Sussex, UK.

[32]Metalsky, G. I., Joiner, T. E., Hardin, T. S., and Abramson, L. Y. (1993) Depressive reactions to failure in a natural setting: a test of the hopelessness and self-esteem theories of depression. *J. Abnormal Psychol.* **102,** 101–109.

[33]Rachman, S. (1990) *Fear and Courage, 2nd ed.,* Freeman, New York.

[34]Kluge, E. W. (1992) *Biomedical Ethics in a Canadian Context,* Prentice-Hall Canada, Scarborough, Ontario.

[35]Munson, R. (1999) *Intervention and Reflection: Basic Issues in Medical Ethics.* Nelson/Thomson Learning, Scarborough, Ontario.

[36]LaFollette, H. (2000) *Ethical Theory,* Basil Blackwell, Oxford.

[37]Macquarrie, J. (1972) *Existentialism,* Penguin Books, New York.

[38]Ross, S. and Malloy, D. C. (1999) *Biomedical Ethics for Health Care Professionals: Concepts and Cases.* Thompson Educational, Toronto.

[39]Heidegger, M. (1962) *Being and Time* Macquarrie, J. and Robinson, E. transl. Harper & Row, New York.

[40]Kierkegaard, S. (1975) The first existentialist, in *Existentialism from Dostoevsky to Sartre,* Kaufmann, W., ed., Times Mirror, New York, pp. 83–121.

[41]May, R. & Yalom, I. (1984) Existential psychotherapy, in *Current psychotherapies,* 3rd ed. Corsini, R., ed., F. E. Peacock, Itasca, IL.

[42]Hynde, G. W., Hern, K. L., Novey, E. S., Eliopulos, D., Marshall, R., Gonzalez, J. J., et al. (1993) Attention-deficit hyperactivity disorder and assymetry of the caudate nucleus. *J. Child Neurol.* **8,** 339–347.

[43]Bandura, A. (1977) Self-efficacy: toward a unifying theory of behavioral change. *Psychol. Rev.* **84,** 191–215.

[44]Kierkegaard, S. (1962) *The Present Age* Dru, A. transl. Harper & Row, New York.

[45]Buber, M. (1958) *I and Thou* (transl. R.G. Smith) T&T Clark, Edinburgh.
[46]Heidegger, M. (1966) *Discourse on Thinking,* Harper Torchbooks, New York.
[47]McMullen, L. M. (1995) Newt, tammy, and the task force: what happened to common sense and moderation? *Canadian Clinical Psychologist* **6,** 5–7.

Abstract

Many mental health advocates harbor misgivings about publicly funded managed care for behavioral health services. As states and counties increasingly shift to managed care arrangements, these advocates view managed care cynically as just another attempt to justify budget cuts and reduce access to needed resources. As a consequence, they argue that the potential for great harm exists for behavioral health consumers in a managed care system, particularly for consumers with multiple, complex chronic needs.[1]

Conversely, a growing minority of mental health advocates share a vision of a publicly funded managed behavioral health care system that succeeds where other public-sector systems have failed. These advocates see managed care as remedy for a dysfunctional public health system that lacks accountability, coordination among government agencies, and a continuum of care for behavioral health consumers. Their vision is premised on a commitment to certain basic, ethical values. "Values-based" publicly funded, managed behavioral health care systems emphasize the individual's recovery while treating consumers as partners in their therapy and rehabilitation. These so-called values-based service plans are driven by consumers' goals and build on consumers' strengths while exercising responsible stewardship of scarce, public resources. In short, such plans strive to provide the right amount of care, at the right time, in the right setting, and for the right reasons.

According to these latter mental health advocates, their vision of public mental health will succeed where others have failed, in part because the delivery system integrates ethics into operations by utilizing a comprehensive corporate ethics program. Such programs have been in use in other industries since the late 1970s. Companies such as IBM, GE, Hewlett Packard, and Levi Straus, Inc. among many other Fortune 500 companies have instituted ethics programs as a means to strengthen their corporate cultures, increase productivity and boost long-term growth, as well as enhance their public image. It is not until quite recently that health service organizations have begun to reap the benefits of ethics programs.

A small but growing number of publicly funded, managed behavioral health care corporations have begun to make use of such programs. For example, Community Behavioral Health, Inc. (CBH), a Philadelphia-based public-sector managed behavioral health care corporation, utilized the services of the Center for Ethics in Health Care, in Atlanta, to implement a systemwide corporate ethics program. The program is largely an attempt to institutionalize the vision of Estelle Richmond, Health Commissioner, city of Philadelphia. This vision is of an integrated behavioral health system that supports a continuum of care. The commissioner's vision includes (1) a commitment to a set of core ethical values, (2) an idea of the mission of public-sector managed care, and (3) a number of important, concrete goals for Community Behavioral Health, Inc. In this chapter, I provide an outline of this vision for public-sector managed behavioral health care. I consider this vision in light of the CBH Ethics Initiative. I discuss how the staff of the Center for Ethics in Health Care, Atlanta worked with management and staff at CBH to integrate ethics into the operations of CBH in an attempt to institutionalize Commissioner Richman's vision.

Managing Values in Managed Behavioral Health Care

A Case Study

Mark E. Meaney

A Vision for Public-Sector Managed Behavioral Health Care

The use of managed care for public-sector behavioral health services is a relatively recent development. Private insurance plans have used health maintenance organizations (HMOs) and other forms of managed care for many years, but public agencies have only recently begun to adopt this approach. In the public sector, managed behavioral health services generally involve contractual arrangements between a public agency, such as a Department of Health, and either for-profit companies or non-profit private entities that agree to "manage" the use of mental health and/or drug and alcohol treatment services. The city of

From: *Biomedical Ethics Reviews: Mental Illness and Public Health Care*
Edited by: J. Humber & R. Almeder © Humana Press Inc., Totowa, NJ

Philadelphia is unique in this regard. Under the direction of Commissioner Richman, the Department of Health created its own managed behavioral health care corporation. Instead of relying on a for-profit company or a nonprofit private entity, the Health Department uses Community Behavior Health, Inc. (CBH) to manage Philadelphia's mental health and addiction treatment services.

This means that CBH contracts directly with a network of providers. Providers in the network receive a capitated rate for each consumer enrolled in the CBH health plan. CBH thus acts as the third-party payer between consumers and their treatment providers. In most instances, primary care physicians authorize payment with CBH for services, in which case, physicians act as "gatekeepers." In other instances, treatment providers must receive prior approval from a CBH case manager before initiating new services or continuing a service beyond a certain point.

By contracting directly with providers, CBH can exercise tight controls over the use and funding of services. Such tight controls, in turn, enhance accountability. With only one stream of funding, there can be no mistake about who is responsible for decisions that impact consumers. CBH can thereby ensure appropriate care and constrain unnecessary use of services. However, along with such tight controls comes an increased risk of potential abuse. Abuse could come in the form of denials of appropriate treatment, coercion of consumers into accepting services they do not want, retaliation against treatment providers who advocate for appropriate levels of care, unqualified case managers who make poor decisions on the need for care, referrals to geographically inaccessible network providers, and retroactive denial of preapproved treatment services.

The vicissitudes of politics make the potential for such abuse particularly worrisome. Changes in public office can impact the operations of public agencies for better or worse. Although the potential for abuse exists in managed care generally, it is particularly onerous in a public health context. Typically, the population in public-sector mental health has more serious disability and mental health needs than the privately insured population. Moreover, poverty and homelessness often compound the needs of public behavioral health consumers. Public mental health systems thus serve as "agents of last resort" for people with no other means of accessing services. Thus, poorly run public-sector managed behavioral health care would reduce the benefits of, and hence deny necessary treatment to, the most vulnerable among us. Changes at the highest levels of government office can thus jeopardize efforts to incorporate the very best of managed care into public mental health. Mental health advocates must try to insulate operations as much as possible from external, political pressures on the delivery of care.

Commissioner Richman is circumspect in regard to the political pressures on the delivery of behavioral health care services. She argues that in the face changing political tides, only a particular philosophical approach to public mental health administration coupled with a vision of managed care can sustain public-sector managed behavioral health care. Her philosophical approach could best be described as a pragmatic idealism. She is idealistic, but with a realistic outlook on the possible. Her philosophy embraces two seemingly contradictory ideas: stability and innovation. Where most public health administrators see an "either/or," Commissioner Richman sees a "both/and."

For Commissioner Richman, an unwavering focus on mental health consumers is the constant. In her mind, public mental health administration ought to be "people centered" and focused strictly on the needs of the individual. Second, however, public health officials must also bear in mind that health care is one social expenditure among many others. Politics consists largely of balancing competing goods in the allocation of scarce public resources. Public mental health administrators must therefore constantly adapt to changes in budgets. Budgetary constraints can have a dramatic impact on the level of resources available to low-income people for health care. Administrators must therefore be innovative in adapting to such constraints. Public health care delivery systems must be efficient as well as effective.

Commissioner Richman's philosophy of public health administration, in turn, grounds her vision of public-sector managed behavioral health care. Traditionally, public-sector behavioral health delivery systems have been anything but systematic. They have been fragmented at best. Diverse streams of parallel funding flowed side by side with no communication among the different elements within the system. Such fragmentation causes waste and ultimately leads to poor quality care, which, in turn, leads to poor outcomes for consumers. Waste of scarce resources also limits the options available to consumers. The limitation of options limits consumers' choices, which, in turn, limits their ability to control their own lives. Conversely, unity among diverse elements within the delivery system increases the options on how best to use scarce public resources. The more options, the better the care; the better the care, the better the outcomes for individual consumers, and the more flexible you can be with the dollars you save.

In short, this vision of public-sector managed behavioral health care is of a system that provides the right kind of care, at the right time, in the right setting, and for the right reasons. First, consumers should be able to access services when they need them. Second, health plans should emphasize care in the least restrictive setting of the consumer's choice. Third, plans should provide a full array of community-support and family-support services, such as psychiatric rehabilitation, case management, preventive treatment, in-home services for children, school-based day treatment, and consumer-run programs. By substituting support services for expensive in-patient and residential services, health plans save money. Moreover, the continuity of care fosters greater efficiencies among providers and better coordination among public mental health agencies. Because a single entity is clinically, financially, and administratively responsible for care, there is a single point of accountability for integrating care and assuring good outcomes. Finally, the vision also involves the application of computer technology to track performance in order to report on the impact of services and outcomes for consumers. A "people first" philosophy of administration and an innovative approach to public-sector managed behavioral health care ensures that no one "falls through the cracks" and the city saves scarce public resources.

The Core Values of CBH

Commissioner Richman's vision of an integrated delivery system led to the idea of Community Behavioral Health, Inc. The narrative of CBH is as much a story about values as it is about systems integration. The story begins with the assemblage of a group of highly dedicated and motivated individuals. The members of the management team were chosen carefully. The stakes were high. The team would be responsible for the creation of the organizational structures that would have to meet the needs of some 400,000 mental health consumers in the greater Philadelphia area. Most team members shared Commissioner Richman's brand of pragmatic idealism (i.e., her "people first" philosophy, her vision of managed behavioral health care, and certain core values).

Commissioner Richman describes the original six members of her team as "like a family." Most believed fervently in a small set of general guiding principles. Although their goal was to make an antiquated mental health delivery system more efficient, efficiency was not the only goal, or even the most important one. Principles came before strategy. Thus, this set of core values united the team to a purpose beyond mere systems integration. They had a commitment to a belief that although public administrations may come and go and although management and employees, strategies and policies, and services and treatment providers may change, these underlying principles of organizational structure must not. The mission was to build an organization that would institutionalize and preserve the vision as well as these underlying core values.

Two principles serve as the cornerstone of CBH. In keeping with the philosophy and vision, the team sought first to promote the best interests of consumers and their families while respecting a consumer's right to self-determination. For Commissioner Richman, consumer choice must be the fundamental value of a publicly funded, managed behavioral health care system. Moreover, administrators and providers must be held accountable for meeting the needs of the people they serve. Thus, the team focused foremost on creating organizational structures that would provide the right kind of care for the individual consumer in a setting of his/her own choice.

This meant that those structures would have to respect the fiduciary responsibilities of treatment providers to individual consumers. The overarching duty of treatment providers to their individual patients is to provide care according to prevailing professional standards. As a consequence, provider and consumer must be able to engage in an ongoing dialogue, in which the parties bring their respective strengths to the treatment decision (i.e., the provider's professional knowledge and the consumer's self-knowledge). The organizational structures that bring provider and consumer together must therefore support this conversation in order to reach the best possible outcome for the consumer. Thus, the team sought to create systems to ensure that consumers have meaningful involvement in the design and delivery of their care. Consumers are involved from the planning stage, so that the plan of care both meets their needs and protects their rights. They receive clear information that describes their rights, the covered services, and how to access services. Consumers ought not to encounter barriers when they need treatment, but when they are dissatisfied with their plan of care, CBH provides them with clear direction on how to find help or to lodge complaints and grievances. In Commissioner Richman's mind, a commitment to core values demands that the delivery system be seamless and invisible.

Along with beneficence and respect for consumer autonomy, the team also sought to create systems that steward scarce public resources for an entire consumer population of 400,000 members well into the future. As a core value, the team institutionalized distributive justice through a variety of means. From the first, team members developed an integrated delivery system with a continuum of care. Where the old system lacked coordination among agencies and treatment providers, resulting in waste and poor outcomes, the new system would unify diverse elements into a single whole. Where the old system focused on providing episodic, emergent care, the new system would ensure that health care professionals would help consumers from diagnosis throughout the course of their treatment.

Values integration has also proven effective in helping CBH control costs through means other than by providing less or inadequate care for consumers. For example, an emphasis on the principle of consumer self-determination promotes consumer, family, and advocate participation in the development of health service plans. Consumers are fully involved in all treatment decisions. They can choose to refuse any treatment that they feel is inappropriate. As a consequence, consumers frequently choose to use inpatient and residential services that emphasize care in the least restrictive setting. Such initiatives substitute less expensive support services for expensive clinical or emergent care. This, in turn, fosters efficiency among treatment providers. Values integration has meant that treatment providers have been linked along a continuum that expands access, maintains a high level of quality, and protects consumer's rights in the most cost-effective manner.

A simple set of core values, then, served to provide substantive guidance in the development of an integrated behavioral health delivery system. Along with a certain philosophy and vision, the principles of beneficence and a respect for consumer autonomy guided team members in the creation of systems that enhanced consumer input into the design and delivery of their care. A commitment to the principles of distributive justice served, in turn, as the basis for the creation of a continuum of care that held providers responsible and accountable.

The CBH Ethics Initiative: Building Corporate Ethical Culture

In the first year, CBH experienced tremendous growth and success. From the original six team members, CBH grew to approximately 200 employees. Along with this tremendous success, however, the organization experienced growing pains. Commissioner Richman and her original team were concerned that new employees might not appreciate or understand CBH's unique approach to public-sector managed behavioral health care. The original team was like a family. And like a family, everyone had an intuitive grasp of principles and values. Team members knew each other intimately. They could count on each other to uphold CBH's philosophy, vision, and core values. At a certain point in its development, however, the organization went from a family-like atmosphere to a legitimate, managed behavioral health care corporation. Commissioner Richman and her team then faced a challenge of a different sort. They faced the challenge of creating and sustaining a strong ethical organizational culture in a growing public mental health corporation. In short, CBH needed a program to promote strong ethical organizational culture and to encourage ethical behavior among all new employees. Commissioner Richman contacted staff at the Center for Ethics in Health Care, Atlanta and launched the CBH Ethics Initiative.

Staff at the Center for Ethics in Health Care had developed a corporate ethics program specifically for health service organizations based on a very simple premise. Every organization has its own distinctive ethos, culture, and value commitments. These unique characteristics derive from its own history and traditions, its own rules and ways of operating, and its own purposes and goals. Every organization also has its own particular place in a web of social and cultural relations. Employees gain insight into ethical development, both as individuals and together as a community, by exploring the ethical values embedded in the culture and history of their own organization. Our job was to help management and staff make explicit and then institutionalize CBH's core values.

The fact that CBH was a new organization proved beneficial. Its original leaders had already imported values, purposes, and goals into its corporate charter. Their efforts gave initial impetus and direction to the growth and development of CBH. Over the course of the first year, management and staff learned to respond to crises, to celebrate the good times as well as endure some bad. In this way, patterns of actions had begun to evolve both for the development of policy and for its implementation. These organizational patterns, or habits, entailed specific value commitments. Staff at the center believed that it was crucial to the health of CBH first to make explicit the value commitments that were embedded within its original charter and policies. The institutionalization of ethics throughout the organization depended on it, as did CBH's own unique identity. The articulation of, and commitment to, values are an essential means by which an organization defines its own notions of corporate integrity.[2]

Events and strong leaders mark the history of an organization. The immediate past of CBH consisted of a chronicle of the actions of a few for the sake of the many. Yet, the past is more than simply what has happened. We knew that the sweat and toil of Commissioner Richman and her original team would live on in the core values of the traditions or cultural heritage of CBH. Their work would continue to exert a powerful influence on the activity of the organization long after they had retired. Thus, history is not just what has happened in the past, but it is also what is remembered. The history of CBH would be documented not only through institutional structure and organizational habit, but it would also be preserved through stories told about the efforts of the original team leaders. These stories would be told over and over as part of the lore of the organization. They would recount impossible barriers, difficult ethical dilemmas, and the moral courage of these visionaries. The culture of every organization is a rich tapestry of vignettes, anecdotes, myths, and legends that convey shared values and beliefs. The corporate ethical development of CBH would involve tapping this rich vein of CBH's "origin narrative" to build commitment to its fundamental moral values at all levels.[3]

Thus, corporate ethical development at CBH would require more than just an ethics committee, a code of conduct, or a set of guidelines and rules, although these too would be important. The institutionalization of value commitments throughout the organization required the following: (1) the development of an ethical culture that rewards ethical behavior, (2) the integration of value commitments throughout operations, and (3) the fusion of ethical reflection into the daily practices of management and staff. We felt that responsibility for corporate ethical development at CBH must extend from the commissioner through senior management, all the way to frontline employees directly in contact with consumers and treatment providers. This goal required communication with, involvement of, and commitment to corporate ethical development at all levels of the organization. How to do this was one of the difficult problems CBH now faced.

For ethical culture to infuse CBH meant, first, that there should be common understandings that unite employees through a sense of community. Open communication is possible when there is an appreciation of a community of purpose. Management and staff at CBH chose to express CBH's community of purpose through a vision statement. It meant, second, that management and staff had to have a strong sense of corporate identity. Although individuals may perform a diversity of tasks, everyone ought to feel like they contribute in different ways to a shared mission. Management and staff chose to make explicit their shared mission in a CBH mission statement. Third, corporate ethical culture also entails a common understanding of organizational integrity. Mere conformity to external guidelines or regulations does not yield organizational integrity. Rather, solidarity based on common ethical values grounds organizational integrity. As value commitments became explicit through a Code of Ethical Conduct and institutionalized throughout CBH, habits of action would reinforce cooperation and coordination beyond the legal minimum. Corporate ethical culture would only then begin to take root at CBH. Fourth, corporate ethical culture means that there is a network of support that sustains participation and innovation. CBH employees should feel empowered to discuss sensitive or controversial ethical issues regarding the delivery of behavioral health care. Most importantly, they should feel confident that they have the support of management in making "hard choices" in their own daily interaction with consumers or treatment providers.[4] Staff chose ethics education programming for all employees at CBH, including the CBH board and senior management, as a means to empowerment.

In the development of corporate ethical culture, it is important also to remember that an organization is a community. As a community, it consists of an organizational context of individual persons and groups, a system of customs, ethical values, and purposes, and a network of actions and interactions. These are the essential building blocks for developing, shaping, and strengthening ethical culture within an organization. To create and sustain ethical culture at CBH, we felt that Commissioner Richman had to (1) tap the ethical resources of persons and groups inside CBH, (2) develop the proper means to make explicit the fundamental values and principles of the organization, and (3) develop and sustain ethical leadership for the future.

An organization is, of course, made up of persons and groups in relation to the purposes of the organization. Employees bring their own natural talents, value commitments, and experiences with them into the setting of the CBH corporate community. Consequently, it was not necessary to start from scratch to develop and sustain corporate ethical culture. The basic materials were already present among the people at work in the organization. Our challenge was to find a way to make use of this wealth of valuable experience. Of course, the individual ethics that people bring with them to a professional setting is not enough and needs shaping through corporate ethical development. Nevertheless, in the development of corporate ethical culture, it is a matter of incorporating the resources already present among employees in order to remain in touch with evolving values. To this end, management and staff elected to create the CBH ethics committee that consisted of members drawn from three levels within the organization: senior management, mid-level management, and staff.

As a community, CBH is also made up of a system of customs, values, and purposes. Customs evolve over time through relationships and actions that make up the social habits within a corporate community. They may consist of various forms of ritual, ceremonies, conferences, celebrations, parties, and fun times that both represent the culture and serve to strengthen the infusion of values throughout the organization. These customs, then, were an important vehicle by which to reinforce CBH's developing ethical culture.

Values represent the goals and standards used by management and staff for judging actions and relations as either better or worse. They provide criteria for evaluation. Over time, they shape and reshape the culture of an organization. The purposes of the organization express values in terms of goals for the future and serve as the basis of change within the organization. Together, customs, values, and purposes form the glue that binds an organization into a community. They come to be shared by management and staff and form the basis of communication, thus providing a network of commitments that inform action. It was therefore vital to the health of CBH to make its ethical values and goals explicit in order to enhance communal relations. When employees of an organization become more aware of the ethical foundations of their activity, they become more aware of themselves as part of a larger whole. As a consequence, they desire to learn and to apply principles in their work in an effort to further the legitimate interests of this greater whole.

Different organizations have devised a variety of means to develop, sustain, and strengthen corporate ethical culture. Some have chosen a more formal approach by relying strictly on guidelines, rules, regulations, and procedures. Others have combined this formal approach with a more informal means of cultivating ethical culture. We suggested that CBH adopt this latter approach. We would assist CBH in establishing an ethics committee, selecting an ethics officer, drafting "ethics literature," and so forth. However, we would also assist CBH with the creation and publication of a more informal "ethics manual," which would serve as the basis of systemwide ethics education. Management and staff recognized the shortcomings of codes of ethical conduct and chose to supplement their ethics literature with a catalog of "hard cases." This manual would amount to a narrative of the history of ethical reasoning at CBH. We felt that corporate ethical development at CBH depended, in part, on tapping the rich vein of the CBH's narrative to build commitment throughout the organization to its fundamental moral values.

In their study of the CBH "ethics literature" and this more informal "ethics manual," employees not only learn the principles and rules of ethical decision making contained in the ethics literature, but they also learn how to apply those principles to "hard cases." The cases were drawn from different departments: Clinical Management, Administrative Management, Human Resources, Credentialing, Financial Management, Network Management, Claims Management, Information Services, and Quality Management.The manual provides examples of concrete instances of how managers and staff in the past have resolved difficult moral dilemmas within each department. New employees thereby gain insight into the corporate ethical culture of CBH. They come to appreciate the common understandings of ethical values that unite everyone in a corporate community. Consequently, new employees develop a strong sense of organizational identity and of corporate integrity. As value commitments have become explicit and institutionalized throughout the CBH, habits of action have reinforced cooperation and coordination among management and staff beyond a legal or formal framework. As a consequence, CBH has developed a network of support that sustains participation and innovation. Ethical culture empowers employees to discuss sensitive or controversial ethical issues. Most importantly, the culture creates confidence in the knowledge that employees have the support of management in making the "hard choices" in the context of their own daily activity with consumers and/or treatment providers.

Moreover, CBH has taken a number of steps with the assistance of the center's staff to ensure the translation of ethical culture into the structures and institutions of the organization. The CBH Code of Ethical Conduct and informal "ethics manuals" provide a framework for the work of the CBH ethics committees. The ethics committee not only serves as a venue for the review of breaches in the code of ethics or ethical culture, but it also functions as a mechanism for the promotion of ethical reflection throughout all levels of the organization. The CBH ethics committee sponsors events, a website, and a newsletter that keeps CBH employees abreast of changes in societal values, behavioral health care ethics, and related legal issues.

Ethics training is mandatory for all employees. The purpose of ethics education programming is to introduce new employees both to the ethical culture of the CBH and to the infrastructure that supports it. The added benefit, of course, is that CBH will create a pool of ethical leadership for the future. Institutional leadership is essential for the promotion and protection of organizational values. Ethics training ensures that the values and purposes of the organization remain at the core of a corporate ethical culture that sustains and enhances trust both within and outside of CBH.

Implications of Values-Based Managed Behavioral Health Care

We have seen that corporate ethical development begins with the formation and dissemination of "ethics literature." Such literature must indicate that the employees of a managed behavioral health corporation intend to act to protect consumer rights and to steward scarce medical resources. Although corporate "ethics literature" serves an important function, I have noted that corporate credos and/or codes of ethical conduct fail to provide employees with insight into both the grounds of moral principles generally and the justification of ethical behavior within the environment of managed behavioral health. Codes, and so forth give little guidance on how to resolve morally ambiguous "hard cases" or new situations. If managed behavioral health care corporations are to institutionalize the aforementioned core values of behavioral health, employees must understand how to apply moral principles to "hard cases" and new situations. Left standing alone, it is doubtful that ethics literature will have a significant effect on behavior. In fact, I have concluded that such corporate ethics literature will become effective only when it is integrated into the culture and operations of a managed behavioral health care corporation.

In more specific terms, then, corporate ethical development for managed behavioral health requires more than just a code of conduct or a set of guidelines and rules. As suggested earlier, the institutionalization of value commitments generally and in behavioral health care specifically requires (1) the development of a corporate ethical culture, (2) the integration of value commitments throughout operations, and (3) the fusion of ethical reflection into the daily practices of all employees. Responsibility for corporate ethical development must extend from CEOs through mid-level supervisors to frontline employees who interact directly with treatment providers and consumers. Again, this goal requires communication with, involvement of, and commitment to organizational ethical development at all levels of the organization.

From what was said earlier, we can conclude that in order to achieve this goal, managed behavioral health care corporations must (1) tap the ethical resources of persons and groups both inside and outside the corporation, (2) develop the means to make explicit the fundamental values and principles of behavioral health care delivery, and (3) develop and sustain ethical leadership for the future of behavioral health care. Corporations in other industries have chosen a variety of means to develop, sustain, and strengthen corporate ethical culture. Some have chosen a more formal approach by relying strictly on codes, guidelines, rules, and regulations. Others have combined this formal approach with a more informal means of cultivating corporate ethical culture. I recommend that managed behavioral health care corporations adopt this latter approach.

In adopting this approach, I suggest that other managed behavioral health care corporations follow the lead of CBH by drawing upon the customs, the value commitments, and the experiences of the persons and groups within their own companies to make explicit the fundamental ethical principles that ought to inform action. Management and staff at CBH have recognized that corporate ethical development involves tapping the vein of an organization's narrative to build commitment throughout the organization to its fundamental moral values. Other managed behavioral health care corporations ought to do the same in protecting the best interests of consumers while allocating scare resources across a consumer population.

A three-step process will complete this task. First, an ethics survey should canvas the value commitments of a cross-section of employees. The survey accomplishes two things. It provides an indication of the extent to which management and staff have an appreciation for the importance of corporate ethical development and ethical culture. Moreover, it provides the raw data for making progress in the development of ethical culture. The survey should inquire about specific events and cases in the delivery of behavioral health care. The questions should be tailored to help management and staff reflect on the ethical content of "hard cases."

Second, once the data are collected and analyzed, those in charge of the ethics initiative could then draw upon the ideas and ethical reflections of this cross-section to compile a sampling of the most difficult ethical dilemmas in the delivery of behavioral health care. They should then formulate a list of specific questions and work up these ethical dilemmas into sample cases. The questions and cases could be categorized according to specific topics that pose particular difficulties in different departments.

In the third step of this process, those in charge of the ethics initiative should gather together the materials from the ethics survey to produce both the "ethics literature" and the "ethics manual." An "ethics manual" would, of course, supplement the ethics literature. The manual would provide a means by which management and staff can learn both the principles and rules of ethical decision making in the delivery of behavioral healthcare and how to apply those principles to "hard cases" involving, for example, treatment providers and consumers. Just as at CBH, the manual would provide examples of concrete instances of how employees have resolved difficult moral dilemmas in the past. New employees would thereby gain insight not only into the ethics literature but also into the ethical culture of a managed behavioral health care corporation.

Recall that for an organization to be infused with professional ethical culture means, first, that there are common understandings that unite members through a sense of community. Open communication about difficult issues is possible because there is an appreciation of a community of purpose. It means, second, that there is a strong sense of corporate identity. Everyone feels like part of the same organization and contributes in different ways to a shared mission. To further this end, a managed behavioral health care corporation can use the results of the three-step process to provide a framework for ethics education programming for the entire organization. The training should begin at the board level and with senior management and extend from mid-level supervisors to frontline employees directly in contact with treatment providers or consumers.

As value commitments become explicit and institutionalized throughout the organization, habits of action will reinforce continued cooperation and coordination among management and staff beyond a mere formal framework. The organization would, in effect, develop a network of support that sustains participation and innovation. Ethical culture will empower employees to discuss sensitive or controversial ethical issues in the delivery of behavioral health care. Most importantly, the culture would create confidence in the knowledge that employees have the support of management in making the "hard choices" in the context of their daily activity.

Management can take a number of steps to build organizational support in order to ensure that ethical culture gets translated into the operations of the organization. Codes of ethical conduct and informal "ethics manuals" provide a framework for ethics committees and ethics officers. As at CBH, ethics committees will not only serve as a venue for the review of breaches in the code of ethics or ethical culture, but they can also function as a mechanism for the promotion of ethical reflection throughout all levels of the organization. For example, as mentioned earlier, the CBH Ethics Committee sponsors a number of events and publishes a website and a newsletter to keep employees abreast of changes in behavioral health care ethics and related legal issues.

Conclusion

Although many mental health advocates harbor misgivings about managed care for behavioral health services, a growing minority of mental health advocates share a vision of a managed behavioral health care system that succeeds where other systems have failed. These advocates see managed care as a remedy for a dysfunctional mental health system that lacks accountability, coordination among treatment providers, and a continuum of care for behavioral health consumers. Their vision is premised on a commitment to certain basic ethical values. So-called "values-based" managed behavioral health care systems emphasize the individual's recovery while treating consumers as partners in their therapy and rehabilitation. As I have suggested, these so-called "values-based" service plans are driven by consumers' goals and build on consumers' strengths while exercising responsible stewardship of scarce resources. In short, such plans strive to provide the right amount of care, at the right time, in the right setting, and for the right reasons.

Commissioner Estelle Richman is an example of these latter mental health advocates. She believes that her vision of public mental health will succeed where others have failed, in part because CBH integrates ethics into operations by utilizing a comprehensive corporate ethics program. Companies in other industries have instituted ethics programs as a means to strengthen their corporate cultures, increase productivity, and boost long-term growth, as well as enhance their public image. CBH too has begun to reap the benefits of ethics programming.

The CBH Ethics Initiative is largely an attempt to institutionalize the vision of Estelle Richmond. This vision is of an integrated behavioral health system that supports a continuum of care. The commissioner's vision includes (1) a commitment to a set of core ethical values and (2) an idea of the mission of public-sector managed behavioral health care. In this chapter, I have provided an outline of this vision of "values-based" public-sector managed behavioral health care. I considered this vision in light of the CBH Ethics Initiative, the CBH ethics literature and ethics manual, and the CBH ethics education programming. I have also discussed how Commissioner Richman used the staff of the Center for Ethics in Health Care to integrate ethics into the operations of CBH in an effort to institutionalize her vision. Commissioner Richman is confident that CBH will succeed where other behavioral health systems have failed in part because of the CBH Ethics Initiative.

References

[1]Jackson, L. G. (1995) *Managing Managed Care for Publicly Financed Mental Health Services,* Bazelon Center for Mental Health Law, Washington, DC, pp.1–5.

[2]Phillips, J. R. and Kennedy, A. A. (1980) Shaping and managing shared values. *Mckinsey Staff Paper* **12,** 4.

[3]Goldberg, M. (1992) Corporate culture and the corporate cult, in *A Virtuous Life in Business*, Williams, O. F. and Houck, J. W., eds., Rowman and Littlefield, Lanham, MD, pp. 29–50.

[4]Phillips, J. R. and Kennedy, A. A., p. 5.

Abstract

Any human artifact has a design, and those artifacts that serve ends we value ought presumptively to be designed to achieve those ends and to achieve them efficiently and effectively. Our system for psychiatric care is such an artifact, although complex and the result of many individual decisions as well as public policy initiatives and other factors, and its very form is being transformed. The traditional Freudian model of open-ended therapy, one-on-one, psychiatrist and patient, is being replaced by a few consultations and medication. The driving force behind this momentous change is not a calculated decision that the ends our psychiatric care ought to achieve can be better achieved through the transformation, although that motivation no doubt plays a causal role. The driving forces are economic: the creation of new drugs that produce some of the effects psychiatric care is intended to produce and the crimping hand of health maintenance organizations that have radically cut access to psychiatric services and made the traditional model unaffordable except for the very wealthy. The transformation in the form of psychiatric care is not without its benefits, but it is a distressing feature of the changes that they are motivated not primarily by a rational consideration of means to ends, but by factors that ought to be secondary in considering any health care system.

The Changing Form of Psychiatric Care

Wade L. Robison

We are in the midst of a transformation in the way in which we conceive of psychiatric care. The change is not just to how we are providing psychiatric care, but also to how we are thinking about the care itself—about what is wrong with the person with the problems, about what a practitioner can do, and about how best to go about proceeding, given the limited resources. Such transformations should be directed by new discoveries in science, by new ways of understanding how problems arise that are amenable to psychiatric care, and, primarily, by a concern to ensure that those afflicted are treated more effectively, with a greater assurance of success and with long-term relief.

A system of health care ought to have ends related to the health of those served by it: that birth and infancy occur in such a way and in such circumstances that one has the highest chance of continuing to live, that those who live have the best preventive care so that they will be least prone to disease and bodily faults, that those who are subject to disease and bodily faults are well taken care of, and that the elderly receive adequate care—all these without significantly harming such other interests as financial

From: *Biomedical Ethics Reviews: Mental Illness and Public Health Care*
Edited by: J. Humber & R. Almeder © Humana Press Inc., Totowa, NJ

well-being. These five features presuppose a sixth, universality of coverage, and seem, *prima facie*, the minimal conditions for an adequate health care system.

Similarly, a system for providing psychiatric care ought to have as its minimal goals the alleviation of the suffering of those with mental afflictions for the long term. That should occur with the least harm to them, and it ought to alleviate not just the symptoms, but change the underlying psychiatric structure that produces those symptoms and, unless changed, may express itself in other harmful symptoms besides those alleviated.

That our health care system lacks the minimal conditions for adequacy goes without argument, but we are also witnessing a change in the form of psychiatric care which seems driven, as is our health care system, by features that, although relevant to delivering psychiatric treatment, should by no means be primary and should certainly not supplant the end of effective and lasting treatment that we ought to be concerned to achieve.

We can come to see how this is so by comparing the traditional model of psychiatric care, that created by Freud, with the new model and then by examining what I call natural social artifacts and how change is effected in the one of interest to us here, the health care system. We shall then provide an example of the new treatment and its effects before turning to summarize what is going wrong.

I am making a complex causal judgment and it needs to be tempered by a cautionary note that, as with any complex system, the effects of any one feature or set of features can at most have effects that are likely, from our standpoint, not certain. In the same way, the complexity of the system is such that one ought to be chary in suggesting solutions to the problems we are about to examine. They are endemic to the health care system as a whole, and their resolutions are anything but easy.

The Traditional Model

We can tell much about a profession by examining jokes and cartoons about it. Engineers are thought to be single-mindedly concerned about mechanisms, for instance. So when a priest and physician are saved from the guillotine because the blade fails to fall, it is no surprise that with lowered head, the third candidate, the engineer, looks up into the mechanism and says, "I see the problem!" Similarly, it tells us much about psychiatrists, or at least about how society thinks or used to think about psychiatry, that cartoons about psychiatrists are instantly recognizable. The psychiatrist is in a cushioned chair with the patient lying on a couch with a padded pillow at one end. We are then witness to the psychiatrist's notes: "This guy is crazy!" Or we are a third party to the dialogue between them as when the psychiatrist asks the chicken on the couch, "Why do *you* think you cross the road?"

We get a picture from these cartoons of what is to "see a psychiatrist" or, more accurately, of our understanding of what it is to see a psychiatrist. We can focus on what is supposed to be the funny bit because the rest is all so familiar. The picture of a psychiatrist sitting in a padded chair, listening to and responding to a patient lying on a padded couch, resonates with our understanding.

First, there are no instruments present—no thermometers, no stethoscopes, no hammers for checking reactions. The psychiatrist is not examining the patient in the way a medical doctor would. Indeed, we can see the difference between cartoons about psychiatrists and those about medical doctors by the paraphernalia in the latter's office—an examination table, the stethoscope around the doctor's neck, the patient partially undressed, and so on. The psychiatrist has no instruments and the patient is fully clothed and does not have a physical problem, a problem with the body. The problem is mental.

Second, in looking in on the psychiatrist at work, we are privy to a methodology that has become the hallmark of our understanding of psychiatry—a psychiatrist querying a patient to tell "the story." When we visit a doctor's office, we are asked for what we think our symptoms are, if any, but our responses are checked by blood samples, soundings of our heart and lungs, and other empirical tests—all directed to determining the state of our body. No such equipment is to be seen in the psychiatrist's office. The methodology seems wholly dependent on the questions of the psychiatrist. That is why the query "Why do *you* think you cross the road?" is funny. The query parodies the unskillful psychiatrist as well as putting the burden on the patient. It is as though the psychiatrist were saying, "How odd! I haven't a clue why you are doing that. You tell me!" How is the chicken supposed to be any more cognizant than any of us of the grounds for such odd behavior? However, the query also tells us the nature of the methodology. It depends on the patient's capacities for self-revelation and self-understanding and on the querying skills of the psychiatrist. The assumption of the methodology is that the patient is to think about his or her own behavior, that there is nothing wrong that a little self-awareness will not fix.

Third, that query also gives us a sense of how open-ended and thus how long the process can be. "Why do you think you cross the road?" is a form of question replicable *ad infinitum.* Every statement the patient makes can be queried by "But why do you think that?" and the questions are as numerous as the expressions of behavior. "Why do *you* think you crow when the sun rises?" So the cartoons carry the sense that, at a minimum, the process is going to take a long time and, worse, that the methodology envisages no natural ending—no final determination of brain failure, mental seizure, or emotional flu. The cartoon says, "This is going to take a long time!"

Fourth, it is as true of the psychiatrist's office as it is of the doctor's office that the professional has the power and the patient the problem. Cartoons turn on that feature as well, as when the psychiatrist says, "Wow! You need professional help!" The patient in a doctor's office is usually shown only partially dressed, subject to all the loss of position and power that entails. The patient in the psychiatrist's office in cartoons is usually shown lying down—although we know that patients generally now sit—and the position emphasizes the patient's passivity in the face of the supposed knowledge of the psychiatrist and the power the psychiatrist has to heal.

These cartoons also imply two more features of our understanding of psychiatry. It is, fifth, labor intensive, and, sixth, it is not for the masses. It is labor intensive because the querying is one-on-one, and because each query leads to an answer that can be queried yet again, or, put another way, because there is no end to the complications a single human being can have, the querying can go on and on. The sorts of problem being investigated tend not to be amenable to ready resolution, in that resolving one of their symptoms often means they will express themselves in new ways that will need to be investigated, one-by-one, one-on-one.

Psychiatry is not for the masses because, as the leather chair and couch tell us, the psychiatrist's time is not cheap. Not everyone can afford psychiatric help on the model of psychiatry we see from these cartoons. It costs money and it takes time, and not everyone has the time or the money for psychiatric treatment. In short, psychiatry as displayed by the cartoons is possible only with certain social conditions. Just as one cannot be a vegetarian in a land of few vegetables (how many eskimos can be vegetarians?), so one cannot have psychiatrists or psychiatric patients in a country without leisure time and a rich leisure class.

It is questionable whether these six features accurately describe all of the features of what was the practice of psychiatry.[1] However, they certainly describe a set of distinctive features that we would all recognize as part of that practice. The nature of the problem being treated, the methodology used to treat the problem with its implications for the long-term nature of the treatment, the power psychiatrists have over patients in that treatment with its labor-intensive nature and their querying mode of operation, and the restriction of treatment to those well heeled enough to afford the time and money—these are all features we recognize as part of psychiatric treatment, and that they are replicated in cartoon after cartoon tells us just how recognizable they are.

A New Model

These sorts of cartoon still occur with great regularity, but new ones are also appearing that present quite a different understanding of psychiatry. The one entitled "If they had had Prozac in the nineteenth century" by Huguette Martel is a delight. It consists of sketches of each of the following famous 19th-century minds:

Karl Marx: "Sure! Capitalism can work out its kinks!"
Friedrich Nietzsche: "Me too mom. I really liked what the priest said about all the little people."
Edgar A. Poe, gazing upon a raven: "Hello Birdie!"

In short, it would be a mistake to think that the six features of psychiatry laid out earlier describe what is presently the general practice, for each of these features is in the process of change.

1. The nature of the problem being treated. It is becoming increasingly assumed that whatever the problem is with a patient (i.e., however it is manifested), the difficulty is some chemical or hormonal imbalance.
2. The methodology used to treat the problem. The methodology for therapy no longer tends to be query and answer. That is how it starts in order to identify the problem, but it stops soon enough with a prescription for medication with only an occasional visit with query and answer to be sure things are going right. If the problem is chemical or hormonal, medication to correct the imbalance is the obvious first choice.
3. The long-term nature of the treatment. The process is no longer open-ended in terms of a continuing therapeutic relationship between the patient and the therapist. It is open-ended in that medication continues, but patients self-medicate.

4. The power psychiatrists have over patients. It would be a mistake to say that the therapist is no longer in the position of power in the relationship. As long as the therapist can stop the prescription for medication, the therapist is in power. However, the aim is increasingly to empower the patient by providing the patient with medication that will allow the patient to regulate his or her own hormonal and chemical states. Is 100 milligrams too much? Try 50. Is that too little? Try 75. These are decisions the patient can make, based on the effects the patient can perceive without necessarily needing a therapist to point them out.
5. The treatment's labor-intensive nature. The therapy is no longer labor-intensive. "Take a pill and see how you feel in the morning!" is a parody of the operating rule because it takes far longer for a regimen of medication to take effect and far longer for a patient to get a sense of what changes have occurred, if any, because of the medication. However, the parody captures an essential point. The therapeutic relationship is now far more like that between a physician and patient in that the former diagnoses and gives medication, the latter takes the medication and determines if the symptoms continue or cease or diminish, and the relationship is no longer open-ended—there being no need for continual therapeutic sessions. Even if the prescribed medication does not work, the end is the same—to find one that does—and the number of sessions needed is limited at least by the end to be achieved.
6. The restriction of treatment. The medication is relatively inexpensive and generally supported by health insurance. So therapy is not just the privilege of those able to afford psychiatrists at $100+ an hour for session after session. Now the masses can afford the few sessions required to get the medication right for long-term relief.

The Causes of Change

The changing landscape of psychiatric care is the result of two very different forces being brought to bear upon the traditional model of how psychiatry ought to proceed:

- On the one hand, the development of new drugs that seem to have the effects therapy hopes to achieve has seemingly made the traditional query-and-answer sessions inefficient and superfluous.
- On the other hand, changes in health insurance and, in particular, the increasing reliance on health maintenance organization (HMOs) with their cost-cutting concerns have forced quicker and less expensive therapeutic solutions. The changing landscape of insurance and health care is changing the very form of treatment.

These two changes have been working in tandem, as we might suppose. It is far less expensive to meet once with a psychiatrist, with perhaps a few follow-up meetings, and take medication for whatever it is that ails us than it is to work through the problems with a psychiatrist over the long period of time that is necessary for us to come to grips with our psychological difficulties and then to work at changing ourselves without causing new problems in the process.

Unfortunately, what is primarily driving these changes, and thus fundamentally altering the psychiatric/patient relationship, is not a judgment about what is the best treatment. What is driving these changes is a judgment about what is the most efficient and least costly way for HMOs to handle the symptoms of psychiatric problems, combined with a judgment about what is most profitable for drug companies to develop, produce, and market.

What is being changed is what I call a natural social artifact, and it is important that we understand both that our health care system, including the forms it imposes on psychiatric care, is such an artifact and that the way in which some such artifacts are created carries no assurance that they will achieve the ends for which we would think they ought to be designed.

What Is Being Changed: A Natural Social Artifact

Our health care system, including our system for responding to and treating psychiatric problems, is a natural social artifact.[2] Each of the terms "artifact," "natural," and "social" carries weight:

- It is an artifact because it is a human creation, a complex produced by innumerable individual, corporate, and public policy decisions.
- It is natural because, like a natural language, natural needs produce effects. Given our vulnerability to disease and infirmities, we would have some system of health care no matter what. Systems are more or less complex and answer more or less to the needs that produce them because of luck, the wealth of the society, the level of education, and so on through a wide variety of causal conditions.
- It is social because, again like a natural language, it regulates the behavior of those within the system. If one wants to be understood, one must speak in such-and-such a way. That "must" is the sign of the regulative feature of language.[3] Not every string of sounds will do and not any string of sounds is the best way to communicate. Similarly, our having a health care system means that only certain treatments are acceptable. Painting the ceilings red is not an acceptable prescription for ridding someone of bad dreams. The system is social because it is normative.

What we have is a complex entity that is affected by a variety of causal factors, is normative for what is and is not permissible in terms of treatment as well as such other relevant features as cost and scope of coverage, and has effects that need not have been intended by any who played a role in creating the structure. Every social artifact is a complex social construction—its internal features themselves embodying values and its use conditioned by them and producing effects with their own values.[4]

We might think that a natural social artifact such as health care ought to be designed to achieve certain ends such as lowering infant mortality, ensuring that all who are ill receive medication, and so on. Our health care system notoriously fails to achieve what might be thought to be the standard ends for a health care system to strive for, and the lesson is that no simple relation exists between the values that lie behind our individual choices, or behind even our public policy choices, and the values reflected in, and furthered by, the system we end up having. We have many examples of complex social structures with unintended features or effects, including features or effects counter to the very ones we attempted to produce.[5] No single vision animates the whole of such complex systems as that we have for psychiatric care; thus, it should be no surprise that it may fail to do or do well what we would suppose such a system should do.

Think of English as a paradigmatic example of a natural language that does well enough in its general aim of allowing us to communicate, but which fails in a variety of ways to do this as well as it could. The various movements to "purify" the spelling or to have spelling reflect the pronunciation are attempts to bend a social artifact into better shape to reflect more reasonably the objects for which it was created or, we come to see, ought to reflect. What drives changes in the language are not decisions about how we might better communicate, but the individual choices of particular speakers whose ways of speaking and pronouncing words get replicated. "Buzz word" becomes a buzz word, well, phrase, for awhile when it is not clear how "buzz" serves to modify "word" to get the sense intended.

Similarly, what is driving the changes in the way in which we think about psychiatric treatment is not a calculated decision about the ends we wish to achieve in such a system and the best means of achieving those ends. What is primarily driving the changes, as I have said, and so fundamentally altering the psychiatric/patient relationship, is a judgment about what is the most efficient and least costly way for HMOs to handle the symptoms of psychiatric problems combined with a judgment about what is most profitable for drug companies to develop, produce, and market. Different causes produce different effects, and so we end up with a natural social artifact with different norms. The "best treatment" becomes Prozac or some other medication that alters the chemical or hormonal balance, and practitioners are to be criticized if they resist the changed form of treatment. They are "old-fashioned," "stuck in their ways," and so forth.

The new norm takes over, and we are all caught up in it without any particular regard for whether the change is for the good and whether, if we were to ask whether we should change the system in that way to achieve better psychiatric care, we would ever think to answer in the affirmative. In short, changes in any natural social artifact may occur for any of a variety of reasons having nothing at all to do with the ends for which we would think the artifact was created. Whether the changes further or set back the ends we think the artifact ought to further is a contingent question, not usually considered in assessing whether or not to institute a change even when that matter is up for assessment.

We can best understand how the process works by examining it in a particular case and seeing how changes in treatment for a particular kind of psychiatric problem, partially based on research but mainly on economic considerations, have consequences that may diminish rather than further the ends a psychiatrist and patient should prefer to see achieved.

An Example

We are all familiar with instances in which we have acted without apparently thinking. I can still recall the fear that flooded into me as I was riding my bicycle down the street as a boy when, passing a row of parked cars, a large dog I had not seen lunged at me, barking furiously, through the open window of one of the cars. I turned away, without thinking—directly into the path of an oncoming car—which stopped in time. Or consider the following story, rich with ethical implications, of

> . . . a woman who drove two hours to Boston to have brunch and spend the day with her boyfriend. During brunch he gave her a present she'd been wanting for months, a hard-to-find art print brought back from Spain. But her delight dissolved the moment she suggested that after brunch they go to a matinee of a movie she'd been wanting to see and her friend stunned her by saying he couldn't spend the day with her because he had softball practice. Hurt and incredulous, she got up in tears, left the cafe, and, on impulse, threw the print in a garbage can.[6]

The current hypothesis from evolutionary biology is that these sorts of reactions, apparently unthinking, are the result of the amygdala kicking in.[7] It stands like a gatekeeper, scanning for trouble, asking, as it were, "Is this something I hate? Something that hurts me? Something I fear?" and then reacting instantly if the answer is yes, sending a message to the brain, triggering "the secretion of the body's fight-or-flight hormones, [mobilizing] the centers for movement, and [activating] the cardiovascular system, the muscles, and the gut."[8] Although the description of how this gatekeeper functions makes it sound deliberative, it is not. The reaction occurs before the mind is even conscious of the experience that has triggered the reaction, experiments show, and the mind is being flooded with hormones before any deliberation can occur. We react, and, in addition, the hormones have their effects. It took me awhile to come down from the fright I received while riding that bicycle.

The evolutionary explanation for the role of the amygdala is that those individuals who deliberated about whether the lion they saw ought to cause them to flee did not pass on their genes. An evolutionary advantage accrues to those whose responses to danger bypass the conscious mind—even if the gain is only the microsecond or two it takes to jump out of the way while one's more deliberative companion is caught.

The sorts of experiences that cause the amygdala to trigger such responses is an object of study, and in addition to the experience of something frightening (a lion, a dog lunging), the suggestion is that our early experiences set up the amygdala for many of our immediate responses—my responding viscerally when I perceive an injustice, particularly one I perceive to be directed at me, one spouse abusing the other over some slight imperceivable as a slight to an objective observer, a person recoiling at the sight of a dog, and so forth.

In each of these cases, the perception to which the amygdala responds has a value—"rough, wordless blueprints for emotional (and moral) life."[9] The prehistoric individuals who, upon perceiving a lion, reacted, without thinking, with "Nice kitty!" did not survive. Thus, one crucial implication of our understanding of the role of the amygdala is that the perceptions "that reach the conscious mind are not neutral perceptions that people then evaluate: they come with a value already tacked onto them by the brain's processing mechanism."[10] We see a lion, the amygdala sets off a response, triggering what the body needs to move with fear, and we then become conscious of the perception, the value the amygdala adds already a part of what we now are conscious of.

We do not always (if ever) perceive and then deliberate, judge, choose, and then act. We (often) perceive and react and then must come to grips with our flooded brain, which may impede, for some time, the sort of rational observation and interpretation of our reactions that we so prize in a moral life. We may thus react wrongly to some perceived slight, see quickly upon becoming conscious of what triggered the amygdala's reaction that the facts do not support that reaction, and yet be unable to respond well to correct our error because our brains are still flooded with the hormones triggered by the amygdala's misperception.[11]

The evidence suggests that we can change (at least some of) the amygdala's reactions, reworking the relatively hard internal wiring that adds one value rather than another and so produces one reaction rather than some other. However, the evidence also suggests that this can be very difficult in certain types of cases,[12] and how we are able to do that, if we can—how the will enters in—looks as though it is going to be a very complicated story.

This sort of analysis of some of our "gut" responses has been used to explain why it is that some men are abusive in certain ways and why some women are ultrasensitive to changes in the emotional climate around them. The hypothesis is that being neglected, or feeling neglected, as a child gave them a need for dependency on others, but a defensive autonomy as well, manifesting itself in "a destructive entitlement," as one person has put it.

It is an odd combination, but deadly to relationships. In women, it tends to manifest itself as a heightened sense of awareness of any emotional change. If her spouse gets out of the wrong side of bed in the morning but, after showering, feels ready to meet the world with a smile, she will be aloof emotionally and back away from intimacy. She will not even be aware she is backing off; the amygdala kicks in for self-defense, and the spouse is left feeling hurt and bewildered at her sudden and apparently uncaused distance. "What did I do now?" he asks and looks in vain for an answer.

In men, it tends to manifest itself in abuse. The man ties himself up with a woman, on whose doting he depends for self-esteem, but when he feels that she gets too close, he backs off because, the hypothesis continues, if she gets too close, she will see how worthless he really is. Again, he is not necessarily aware of any of this behavior. He may not even see his behavior as abusive, let alone as something he is manifesting because of early neglect. The effect is that he claims for himself all of the attention she has to give, and when she gives it, he backs off and then gets angry if she does not continue to give him the attention he needs. The woman feels used and abused and finds it difficult to understand how the man could not know that he is abusing her in this way.

Medication can do wonders for this condition in men, mellowing out the anger that accompanies it so that little is left. However, what is lost with only medication is any change in the fundamentals of the behavior and the conditions for their being manifested. Medication does not make the person feel any greater self-worth, and although it will mellow out the anger, it will not itself change the movements in the "game." The person will still claim exclusive and concentrated attention, back off when it is forthcoming, and get angry, albeit in a more mellow way generally, when it is then not forthcoming.

The medication, in other words, modulates the causal relations, but not the underlying causal mechanism. For that, it seems that what is needed is old-fashioned psychiatric give-and-take, a coming to grips by the individual of the pattern of behavior, of what sets it off, and of how it might be controlled or manifested in different ways.

We can understand opting for medication that has no known significant adverse side effects but that causes a less abusive personality and so allows space for more understanding. But, why would anyone opt for it as the exclusive solution to the problem? Topping off the oil in a car will moderate the symptoms of loss of oil, but do nothing for the underlying causal mechanism that causes the oil to leak or burn. Continuing to top off the oil will continue to control the symptoms, of course, but we have no guarantee that other adverse symptoms that are not so obvious are being moderated and we do have a guarantee that the underlying mechanism has not changed.

Means to Ends

What is driving a transformation in psychiatric care is not a calculated and thoughtful set of decisions about how best to treat patients, but what is the least costly way to control the symptoms. There are gains to this. More people can be provided with the means to alleviate the symptoms they have because the costs of drugs are less than the open-ended psychiatric care in the traditional model. Also, the drugs do work to alleviate the symptoms in many cases where the traditional model too often fails. However, there are losses as well. The behavior gets changed, which is all to the good, but not the sources of the behavior. So the need for medication continues, and the person being treated is no nearer to understanding what makes him or her tick than before starting treatment.

This is generally true of our health care system. Too many decisions are driven by financial rather than medical considerations. In a moving account of his wife's illness, Jesse Lemisch described how the insurance company intervened in his wife's treatment, and he argued that the concept of managed care "enables Metropolitan Life to overrule physicians' judgments".[13] Independently of the truth or falsity of his particular account, a system of health care that displaces the physician–patient relationship and substitutes judgments based on other than health-related considerations is a system in which the physician–patient relationship is no longer primary. A health care system will always be affected by other than health related concerns, but such examples imply that other considerations have become primary too often and are entering the system at particularly ill-chosen spots in the intervention process regarding ill-chosen procedures, having particularly bad effects on those who are ill.

The most striking example of ill-considered entry of non-therapeutic considerations is in therapy where "a system has been put in place that has, in effect, interposed a third party into the previously sacrosanct, confidential relationship between doctor or therapist and mental health patient."[14] Such a development fundamentally alters psychiatric practice because its forms will be driven by financial considerations, not success in its treatment of patients. "The most endangered species is the long-term, open-ended core of psychoanalysis that was the norm for decades after Freud developed the field."[15] Whatever the benefits of trying to keep costs down within the current structure of the health care system, managed care puts individual "medical decisions in the hands of bureaucrats,"[16] intervening in the physician–patient relationship and so risking harm to patients, and even threatens to transform, as in psychiatry, the very forms of medical practice, making them primarily responsive to financial rather than medical considerations. Many HMOs and health insurance systems now significantly limit the number of visits per year to a psychiatrist, making long-term psychiatric intervention impossible even when thought essential.

This form of financial intervention is encouraged by the proliferation of drugs that moderate symptoms without apparent ill effects. They are not particularly costly, and they produce some of the desired effects of open-ended psychiatric care of the traditional sort much more quickly. Calming a massive temper is no small accomplishment, and we should not take this criticism amiss. It is not that the drugs are misused or abused, although perhaps they are, and it is not that they do not have good effects, which they do. Rather, the danger is that the form of psychiatric treatment is determined by financial considerations, which elevate the taking of drugs in order to moderate symptoms above the care deemed necessary by psychiatrists to alter completely the sources of the behavior and provide lasting change beyond that produced by medication. We risk harming the patient. We risk harming the profession of psychiatry and the relationship psychiatrists have to their patients. Also, we risk these harms because we fail to ask the simplest of questions, "How can we best design a system of psychiatric care to produce most efficiently and effectively the benefits for the patients?"

Notes and References

[1]I do not mean to imply that these features are exhaustive of psychiatric practice, only that they are complete enough to give us an understanding of how it differs from, say, medical practice and of what some of its more essential features are.

[2]For a more extended, but still brief analysis of the concept, *see* Robison, W. L. (1995) *Decisions in Doubt: The Environment and Public Policy,* University Press of New England, Hanover, NH, chap. 2.

[3]Regarding this *see*, for example, Cavell, S. (1969) *Must We Mean What We Say?* Scribner, New York, pp. 1–43.

[4]For depressing examples of how, for example, commercial pressures can alter even natural phenomena, consider how fruits and vegetables have been altered by their commercialization. The commercialization affects not only the varieties available but also the characteristics of those varieties and the research to produce those characteristics. For a delightful analysis of this process in regard to tomatoes, including a description of the MH-1, a tomato that will rupture only after hitting an impact speed greater than 13.4 miles per hour, "more than two and a half times the speed which federal auto-bumper safety standards [used to] provide for the minimum safety of current models," *see* Whiteside, T. (1977) A Reporter at large: tomatoes, *The New Yorker* **52,** January 24, 36–61; quoted remarks from p. 61.

[5]Naturalists have long thought it appropriate to stock streams with farm-bred trout. The demand for such stocking comes primarily from those who fish and those who profit from selling to those who fish, but the program seems innocent in any case. What can be the harm in artificially increasing the number and quality of trout in a stream?

One difficulty is that such fish are unaccustomed to the warning signs—spreading of fins, for instance—of native trout and the two groups exhausted themselves in territorial fights. The native trout are reduced in number, overcome by the initially larger hatchery-raised trout, but the latter, raised in the ease of life of trout farms, are not subjected to the winnowing process that occurs in a natural environment and they fall prey much more readily than native trout to disease and "stupid" mistakes. Thus, they "fed inefficiently, expended too much energy, grew thinner and often died" [Stevens, W. K. (1991) Hatched and wild fish: clash of cultures, *New York Times*, p. B5]. The hatchery-bred trout *replace*, rather than supplement, the native trout and they then die off more rapidly.

Thus, stocking streams with farm-bred trout is actually counterproductive, producing fewer and less healthy trout in streams than would have existed without the stocking. Similar problems occur when non-native fish are introduced into a new habitat. In California, the introduction of non-native brown trout has decimated the native gold trout [Luoma, J. R. (1992) Boon to anglers turns into a disaster for lakes and streams, *New York Times* November 17, p. C4].

The lesson here is that trying to change a complex social system is much more difficult than one might think. If we cannot successfully alter the balance of fish in a trout stream, but actually do harm, it is not obvious that we can succeed in altering our health care of psychiatric care systems in any easy way.

Similar examples of how readily we can produce the opposite of what we intend abound. Nothing could seem more innocent than putting up nesting boxes for wild wood ducks, but it is a policy that serves to reduce the wood duck population when, as has happened, the boxes are put in places too easily spotted by other wood ducks. The females are as opportunistic as cowbirds and, when they can, lay their eggs in the nests of other wood ducks. Some nesting boxes have been found with as many as 50 eggs laid in them—each a loss [Anon. (1992) Birdhouses for ducks may harm breeding, *New York Times* May 19, p. C4].

Farmers in Florida have discovered that a fungicide, Benlate, used to control molds and fungi to protect millions of dollars worth of crops so poisons the soil that "even weeds will not grow. Cucumber seeds will not germinate. Broccoli plants wither and die" [Raver, A. (1992) Farmers worried as a chemical friend turns foe, *New York Times* February 24, pp. A1 and 14].

[6]Goleman, D. (1995) *Emotional Intelligence*, Bantam Books, New York, p. 16.

[7]The material that follows is drawn from a great deal of reading on the problem of abusive behavior and its causes. Much of the literature is of the popular self-help variety and is painful reading for anyone interested in the theory behind the current research or even in the current research. Among the best of the genre, despite the title, is Terrence Real's *I Don't Want to Talk About It* (Scribner, New York; 1997). Among the more theoretically interesting works are Alice Miller's *For Your Own Good: Hidden Cruelty in Child-Rearing and the Roots of Violence* (Noonday, New York; 1983) and Robert Kegan's *The Evolving Self: Problem and Process in Human Development* (Harvard University Press, Cambridge, MA, 1982).

[8]Real, T. *I Don't Want to Talk About It*, p. 16.

[9]Ibid., p. 22.

[10]Goldman, D (1995) Brain may tag all perceptions with a value. *New York Times* August 8, pp. C1 and C10.

[11]Our problem with coming to grips with what we have done—are doing in the sense that our flooded brain is still supporting our original response—is compounded by what ought to be the reaction of the person we have just unjustly harmed, for that person ought to feel indignant and we must try to atone for what we have just done while also trying to overcome the messages our flooded brain is sending to us.

[12]Bargh, J. A., Chen, M. and Burrows, L. (1996) Automaticity of social behavior: direct effects of trait construct and stereotype activation on action. *J. Personality and Social Psychology* **71(2),** 241.

[13]Lemisch, J. (1992) Do they want my wife to die? *New York Times* April 15, p. A27.

[14]Goleman, D. (1991) Battle of insurers vs. Therapists. *New York Times* October 24, p. D1.

[15]Ibid., p. D9.

[16]Anon. (1992) California's medical model *New York Times* February 17, p. A16.

Abstract

This chapter concerns the balancing of individual privacy interests with societal protection interests. First, I examine Tarasoff *laws, which address the duty of psychotherapists to warn and protect those who might be potential victims of their clients, even though such a duty violates the client's rights to confidentiality and privacy. Although these rulings have a solid foundation in John Stuart Mill's harm principle, they fail to meet several of the conditions upon which Mill insists. First, Mill claims that the threat of harm must be to a specific and readily identifiable person. An abstract threat to others in general does not justify interference with the liberty of an individual. Second, the harm should be reasonably predictable. A vague suspicion that harm might be done is not enough. Third, additional obligations may arise in the context of certain special relationships. If a relationship between two parties, based on certain expectations, has particular consequences for others, both parties have obligations to those others. To enhance practicability, I add a condition of consistency in content and application. It is not appropriate to have conflicting laws in various areas of the United States. To remedy the deficiencies of current* Tarasoff *laws, I suggest that they be modified to more closely resemble Megan's Law, which addresses the duty of law enforcement agencies to notify potential victims of convicted sex offenders. Megan's Law provides a good model because, although also grounded in Mill's harm principle, it meets his additional conditions to justify state interference with the liberty interests of the individual. I conclude, however, with a cautionary note that some recent events indicate a disturbing trend in which Megan's Law is being expanded beyond its legitimate scope of exercise. The law goes too far when the consequences of its application themselves foster harm to others.*

Tarasoff, Megan, and Mill

Preventing Harm to Others

Pam R. Sailors

> . . . everyone would surely agree that if a friend has deposited weapons with you when he was sane, and he asks for them when he is out of his mind, you should not return them. The man who returns them is not doing right, nor is one who is willing to tell the whole truth to a man in such a state.[1]

On August 20, 1969, Prosenjit Poddar told his psychologist, Dr. Lawrence Moore, that he intended to kill a girl. Although he did not identify the girl by name, his comments made it reasonably clear that she was Tatiana Tarasoff, who was spending the summer in Brazil at the time. Dr. Moore, after consulting with colleagues, decided that Poddar should be involuntarily committed to a mental hospital for observation. Moore notified campus police and requested that they detain Poddar. Campus police officers took Poddar into custody, but when he seemed rational, they released him after getting his promise that he would stay

From: *Biomedical Ethics Reviews: Mental Illness and Public Health Care*
Edited by: J. Humber & R. Almeder © Humana Press Inc., Totowa, NJ

away from Tatiana. Seven days later, on October 27, 1969, Prosenjit Poddar killed Tatiana Tarasoff.

In a suit brought by the girl's family, the California Supreme Court, juggling concerns of confidentiality and public protection, held that therapists have a duty to warn potential victims of dangerous patients. Confused by the ruling, the American Psychiatric Association asked the court for clarification. The court reheard the case and, in 1976, replaced its earlier ruling, extending the responsibility of therapists beyond warning, to include protection of potential victims. Now known as the "Tarasoff Principle," the Court concluded:

> When a psychotherapist determines, or pursuant to the standards of his profession should determine, that his patient presents a serious danger of violence to another, he incurs an obligation to use reasonable care to protect the intended victim against such danger. The discharge of this duty may require the therapist to take one or more of various steps, depending on the nature of the case. Thus it may call for him to warn the intended victim or others likely to apprise the victim of the danger, to notify the police, or to take whatever other steps are reasonably necessary under the circumstances.[2]

Although this decision applied only to California, its basic conclusion has been incorporated into law in most of the United States.

The question raised by the *Tarasoff* case and those that have followed elsewhere is how best to balance confidentiality and public safety. More generally, it is the question of when it is legitimate to interfere with the liberty of others. This is not, of course, a new question. One hears echoes of John Stuart Mill in the court's pronouncement that "The protective privilege ends where the public peril begins."[3] Mill's harm principle, from *On Liberty*, wields its influence in widespread areas. It almost seems a given now to assume that one ought to make decisions based on considerations of likely harms to those affected by the decisions.

Mill's famous "very simple principle" tells us:

> that the sole end for which mankind are warranted, individually or collectively, in interfering with the liberty of action of any of their number, is self-protection. That the only purpose for which power can be rightfully exercised over any member of a civilised community, against his will, is to prevent harm to others.[4]

Further, the privilege itself may be limited by illness: "Those who are still in a state to require being taken care of by others, must be protected against their own actions as well as against external injury."[5] So, even if we could make a case against interference based on lack of harm to others, we can, in the case of mental illness, still limit the liberty of one whose illness prevents rational decision making. "Although Mill does not specifically mention the mentally ill, there can be no doubt that he would be horrified to be cited as a source for the doctrine that society has no right to intervene on their behalf. *For it is precisely the reflecting faculty that is diseased in the mentally ill.*"[6]

I will assume, however, a situation in which a person shows no significant lack of rational capacity; that is, although the person's thought processes may be disturbed to a degree by mental illness, competence remains. I will further assume that the words or actions of this person elicit suspicion that he or she could pose a threat to others. Given this situation, Mill lays down three conditions for interference with the person's liberty.

First, Mill claims that the harm must be to a specific and readily identifiable person. An abstract threat of harm to others in general does not justify interference with the liberty of an individual. Some states have incorporated this "specificity rule" in their *Tarasoff*-type laws, ruling that: "If a patient expresses a specific threat to harm another person who is identifiable or reasonably identifiable, then the duty to warn/protect arises, but not unless these conditions are satisfied."[7] Mill's version of this is found in his discussion of whether one can be held responsible for omissions as well as actions. According to Mill, there are circumstances in which I could harm you by refusing to act, and I am just as responsible for those harms as for harms that result from my actions. We are to exercise more caution when assigning blame for inaction, however, doing so only if the relevant facts reveal an obvious and specific threat of harm from the inaction.

> To make any one answerable for doing evil to others, is the rule; to make him answerable for not preventing evil, is, comparatively speaking, the exception. Yet there are many cases clear enough and grave enough to justify that exception.[8]

In other words, it is legitimate to hold one responsible for not preventing harm so long as one ought to have been able to recognize the clarity and gravity of the threat (i.e., so long as there was a serious threat to a reasonably identifiable victim). After all, for the threat to be clear and grave, we must have knowledge of the identity of the person threatened.

Mill's second condition is that the harm should be reasonably predictable. A vague suspicion that harm might be done is not enough. Mill uses the sale of poisons as an example when he outlines how the harm principle should be applied. Poisons may be used for rather banal purposes or for nefarious ones. Because we know that they may be used to harm others, perhaps we should ban their sale in order to prevent the harm that may be done by their ill use. Mill agrees that "it is one of the undisputed functions of government to take precautions against crime before it has been committed, as well as to detect and punish it afterwards." However he reminds us that what is true of poisons is true of much; that is, harm can result from the most innocent things if combined with evil intent. Thus, more caution must be used in preventing than in punishing harm. "Nevertheless, if a public authority, or even a private person, sees any one evidently preparing to commit a crime, they are not bound to look on inactive until the crime is committed, but may interfere to prevent it."[9]

I take Mill's point here to be that it is legitimate to prevent one from failing to act if that failure will result in harm, but we should proceed to do so only with great caution. Unless a reliable prediction can be made, it is illegitimate to interfere with the liberty. Indeed, the prediction should be on the level of seeing that it is evident that a crime will be committed or harm done.

Mill's third condition for interference notes that additional obligations may arise in the context of certain special relationships. If a relationship between two parties, based on certain expectations, has particular consequences for others, both parties have obligations to those others. I may enter into a relationship with you by explicit agreement or by behaving in a way that indicates an intention to continue to act thusly. In either case, upon forming the relationship with you, I incur new obligations toward you and certain responsibilities to you. Further, our relationship may have effects for others, extending our responsibilities to them. As Mill puts it:

> If the relation between two contracting parties has been followed by consequences to others; if it has placed third parties in any peculiar position . . . obligations arise on the part of both the contracting parties toward those third persons.[10]

Mill's main concern in this section seems to have been the welfare of children in case of the dissolution of the relationship of marriage. His description, however, works equally well for the relation between therapist and patient, which may create duties to a third party who may be affected by what occurs in the therapeutic relationship. The therapist enters into a relationship with the patient by which the therapist assumes obligations to the patient. During the relationship, information may be given that has consequences for others outside of the therapeutic relationship. Should this occur, the patient *and* the therapist have obligations to those third parties, obligations that may well require the therapist to act to prevent harm being done to a third party by the patient.

The court in the original *Tarasoff* ruling also noted the necessity of the presence of a special relationship—between therapist and patient—for the presence of a custodial responsibility to warn on the part of a therapist.

> . . . when the avoidance of foreseeable harm requires a defendant to control the conduct of another person, or to warn of such conduct, the common law has traditionally imposed liability only if the defendant bears some special relationship to the dangerous person or to the potential victim. . . . the relationship between a therapist and his patient satisfies this requirement . . .
>
> . . . as explained in section 315 of the Restatement Second of Torts, a duty of care may arise from either "(a) a special relation . . . between the actor and the third person which imposes a duty upon the actor to control the third person's conduct, or (b) a special relationship . . . between the actor and the other which gives to the other a right of protection."[11]

The (a) condition here describes the relationship between the therapist and the potential victim, which creates a responsibility on the part of the therapist to warn and protect that person, even though this will almost certainly disrupt and constrain the potential victim's life. The (b) condition could apply to either of two relationships. If read in one way, it asserts, as in (a), a relationship between the therapist and the person who might be harmed by the actions of the patient. This relation would create a responsibility of the therapist to protect the potential victim. Read another way, the condition describes the relationship between the therapist and the patient, in which the therapist has an obligation to protect the patient from acting in a way that would be injurious to his or her own interests.

Finally, to enhance practicability, I go beyond Mill to add a condition of consistency in content and application. It is not appropriate to have conflicting laws in various areas of the United States. Yet, this is currently the case with laws addressing the responsibility of therapists to warn and protect those who might be harmed by patients. Some courts have held, with Mill's first condition, that therapists have the duty only if there is a specific threat against a reasonably identifiable victim. Other courts have held therapists liable even without a specific threat or a reasonably identifiable victim. Some courts have ruled that the therapist has no duty at all unless the patient is hospitalized. Others have ruled that the therapist's primary duty is to the patient, such that confidentiality cannot be breached even to warn a threatened third party. The Supreme Court in at least one state has gone so far as to extend the therapist's duty to protect harm to property as well.[12] Further, "once a federal or state supreme court adopts a position on this issue, there is no assurance that the same court will not later modify, expand, or restrict its stand in a subsequent case with different circumstances."[13] The standard is thus not only variable from state to state but also likely to shift with changing circumstances within a state.

Although it is true that there are many laws that vary from state to state, it ought not to be the case unless the benefits of variance outweigh the costs of uniformity. In the case of *Tarasoff* laws, there are no obvious or necessary benefits of allowing individual states to determine the scope and conditions of the laws. On the one hand, enhancement of the rights of states is not self-evidently inherently valuable and, in some cases, may run afoul of the national Constitution. On the other hand, the negative results of confusion in the minds of therapists and law enforcement officers argue for consistency in content and application of the laws.

To eliminate the deficiencies of current *Tarasoff* laws, they should be modified to resemble more closely Megan's Law, which addresses the duty of law enforcement agencies to notify potential victims of convicted sex offenders. This law, although also grounded in Mill's harm principle, meets the additional conditions of specificity, predictability, special relationship, and consistency. The law, as is often the case, arose in response to tragedy.

In February of 1988, Jesse Timmendequas was released from prison after serving a 6-year sentence for sex offense against a minor. On July 29, 1994, he raped and murdered 7-year-old Megan Kanka, his neighbor, after luring her into his house with the promise of seeing a new puppy. Maureen Kanka, Megan's mother, claimed that she would have acted to ensure that her daughter had no contact with Timmendequas if she had known of his previous conviction. In response to the crime, the state enacted a law requiring creation of a state registration requirement for all sexually violent offenders. In 1996, President Clinton signed Public Law Number 104-145, better known as "Megan's Law," to require the release of relevant information necessary to protect the public from sexually violent offenders.[14]

Megan's Law could have been written with Mill's harm principle as a guide. Because the law has its origin and longest history of application in the state of New Jersey, all specific references here are to that state's code. New Jersey's guidelines begin with an explanation of the positive consequences expected to result from implementation of the law:

> . . . the purpose of this legislation is to provide pertinent information to law enforcement and, in appropriate circumstances, to neighbors, parents and children, as well as community organizations which care for or supervise women or children. It is hoped that, armed with knowledge of the descriptions and whereabouts of sex offenders and pedophiles, community members will be in the best possible position to protect their children and themselves.[15]

What the law will do, we are told, is inform people who might be affected by the presence of a sex offender. What justifies the law is the belief that such information will enhance the prevention of harm. The Millean reading of this is that we interfere with the free actions of the individual, but only because those actions are likely to affect others; that is, "the conduct from which it is desired to deter him, must be calculated to produce evil to someone else."[16] Looking at the details of the legislation shows that the four additional conditions examined earlier are also satisfied.

Megan's Law meets the specificity rule in several ways. First, the scope of notification is limited to those who are likely to encounter the offender. If the offender is rated as a low risk of reoffense, only local law enforcement personnel receive information about the offender. If the offender is rated as a moderate risk of reoffense, all educational institutions, licensed day care centers, summer camps, and community organizations that are likely to encounter the offender also receive notification. Only when the offender is rated as a high risk of reoffense are all members of the community who are likely to encounter the offender notified. In addition, the law is not static; that is, one's status is periodically reviewed and changed if circumstances warrant.

> Prosecutors and local law enforcement agencies should be cognizant that the determination as to which tier is appropriate in any given case will be an ongoing process. Change of address or information which provides evidence of a change in circumstances or in the relevant factors may trigger a reevaluation.[17]

Thus, the system is crafted such that notification is given only to reasonably identifiable victims or those who might protect them, and the specificity condition is met.[18]

Megan's Law meets the predictability condition with the use of a detailed, specific, quantified measurement instrument, the Registrant Risk Assessment Scale (RRAS). The RRAS examines two general areas to assess the potential risk of reoffense. The *seriousness* of the possible offense and the *likelihood* that there will be a repeat offense are taken into consideration by examination of the following criteria: degree of force, degree of contact, age of victim (mirroring statutory age levels), victim selection, number of offenses/victims, duration of offensive behavior, length of time since last offense (while at risk, not while in jail or civilly committed), history of antisocial acts, response to treatment, substance abuse, therapeutic support, residential support, and employment/educational stability.[19]

After each of these criteria are examined and assigned a prescribed numerical value, it is translated into a risk assessment that corresponds to one of the three tiers. In only two situations can prosecutors modify the tier assignment of the offender. If the offender asserts that he will reoffend if given the opportunity and there is some evidence to support this assertion, he must be placed in the high-risk tier, regardless of how he scores on the RRAS. If the offender has some physical condition (including advanced age or serious illness) which would make him unlikely to reoffend, he must be placed in the low-risk tier, without regard for his score on the RRAS.[20] The detailed and strict guidelines consider and quantify all the various factors that have been shown to aid in predicting the likelihood and severity of future offenses, such that the predictability condition is met.

One way that Megan's Law meets the special relationship condition is by guaranteeing notification of the offender before notification is given. This serves two purposes: It guarantees the rights of the individual and it acts as a safeguard against mistakes. A formal procedure is in place if the offender wishes to object to either the tier placement or to the initial identification. Thus, there is a special relationship between the offender and the authorities that obligates the authorities to provide the offender information and opportunity to challenge the proposed notification. There is also a special relationship between the offender and those likely to be encountered. This relationship is acknowledged by the requirement that the latter parties be notified of the presence of the former. People who are not likely to encounter the offender, and thus not participants in the special relationship, are not to receive notification; in fact, the law explicitly forbids such wider dissemination.[21]

Finally, Megan's Law meets my consistency requirement simply by being federal law. As such, it is equally applicable in all areas of the United States. Granted, there is some variance in how the law is implemented among states, but we are far removed here from the hodge-podge approach seen in *Tarasoff* laws. General guidelines are provided and, in all cases, the purpose of the law is to allow (or more accurately, to require) interference to prevent harm.

I have shown that Megan's Law is well crafted by the test of Mill's criteria for interference with individual liberty, but I must add a cautionary note. Some recent events indicate a disturbing trend in which Megan's Law is being expanded beyond its legitimate scope of exercise. When the Supreme Court of New Jersey ruled that widespread publicizing of the names of sex offenders violated the offender's right to privacy guaranteed by the state's Constitution, citizens of New Jersey voted to amend their Constitution. Now, the Constitution allows not only the names but also the addresses and photographs of sex offenders to be posted on the Internet. Because a federal appeals panel ruled that the law might also violate the national Constitution, the US Supreme Court has agreed to review the law soon.[22]

Should one wonder at the probable consequences of easy access to information about sex offenders, evidence is available from Britain. After *The News of the World* tabloid ran pictures of sex offenders, police noted a sudden and precipitous increase in vigilante attacks.

> Across Britain, several people who shared surnames with alleged offenders named by the paper . . . had their homes attacked. . . . In Portsmouth, England, 70 miles southwest of

> London, police said five innocent families have so far been forced to leave their homes after threats from neighbors. . . . One man reportedly was suspected of pedophilia simply because he lived alone and talked about how much he loved his mother.[23]

Of course, I do not want to condone vigilante action, nor would Mill. Megan's Law is justified by the harm principle. The harm principle is justified by the overall benefits that accrue from preventing harmful actions. It would be self-defeating if its exercise itself *created* harmful actions. Thus, Megan's Law should not be expanded to allow the widespread dissemination of information about sex offenders. The law goes too far when the consequences of its application themselves foster harm to others.

I have shown how John Stuart Mill's work on liberty may be used to provide conditions for interference with the actions of individuals. Of course, interference requires that there be harm to others. In addition, there must be a specific threat to a reasonably identifiable victim. Reliable predictions of harm must be attainable. A special relationship between the potential harmer and the harmed must be present. Also, there must be consistency in content and application of the standard. After establishing these criteria, I applied them to point to deficiencies in *Tarasoff* laws. In the way of contrast, I used Megan's Law as illustrative of a code that meets Mill's conditions. Finally, I concluded with a cautionary note that Megan's Law is being expanded in some areas beyond what is justified by Millean principles. Whenever the law results in more rather than less harm—or, as Mill would have it, more pain than pleasure—the initial justification for interference with individual liberty is lost. Up to that point, however, so long as the four conditions discussed here are met, prevention of harm justifies much. My conclusion is that *Tarasoff* laws should be modified and standardized to more closely resemble the circumscribed form of Megan's Law.

Notes and References

[1]Plato. (1974) *The Republic*, G. M. A. Grube, tranl. Hackett, Indianapolis, IL, p. 5.

[2]*Tarasoff v. Regents of the University of California* (1976) et al., 551 P 2d 334.

[3]*Tarasoff v. Regents of the University of California*

[4]Mill, J. S. (1989) *On Liberty*, Collini, S., ed. Cambridge University Press, New York, p. 13.

[5]Mill, J. S., p. 13.

[6]Isaac, R. J. and Armat, V. C. (1990) *Madness in the Streets: How Psychiatry and the Law Abandoned the Mentally Ill*, The Free Press, New York, p. 338.

[7] Felthous, A. R. (1989) *The Psychotherapist's Duty to Warm or Protect*, Charles C. Thomas, Springfield, IL, p. 23.

[8]Mill, J. S., p. 14.

[9]Mill, J. S., p. 96.

[10]Mill, J. S., pp. 103–104.

[11]*Tarasoff v. Regents of the University of California*

[12]*See* Stone, A. A. (1986) Vermont adopts *Tarasoff:* a real barn-burner, *Am. J. Psychiatry* **143(3),** 352–355.

[13]Felthous, A. R., p. 113.

[14]It may be salient to note that pedophilia *is* categorized as a form of mental illness. *See* American Psychiatric Association (1994) *Diagnostic and Statistical Manual of Mental Disorders,* 4th ed., American Psychiatric Association, Washington, DC, p. 302.2.

[15]Farmer, J. J., Jr. (2000) Attorney General Guidelines for Law Enforcement for the Implementation of Sex Offender Registration and Community notification Laws (revised March 2000), p. 1.

[16]Mill, J. S., p. 13.

[17]Farmer, J. J., Jr., p. 49.

[18]One might wonder about the apparent lack of a reasonably identifiable individual victim in this case. The Registrant Risk Assessment Scale on each offender examines the characteristics of previous offenses. Then, each identifiable group is considered as a potential victim. If the offender preys upon children, then all members of the group "children" constitute the reasonably identifiable victim.

[19]Farmer, J. J., Jr., Exhibit E, pp. 5–10.
[20]Farmer, J. J., Jr., Exhibit E, p. 1.
[21]Framer, J. J., Jr., p. 31.
[22]Newman, M. (2000) Naming sex offenders on the Internet passes in New Jersey, *The New York Times,* November 8.
[23]Byrne, C. (2000) Confused UK vigilantes target doctor, The Associated Press, August 30.

Index

N

O

P